WASTED LIVES OF UNSUNG HEROES

WASTED LIVES OF UNSUNG HEROES

Gripping Stories Of How Czechoslovakian
Communism Persecuted Their WWII Heroes

```
Vítek Formánek
Eva Csölleová
```

SASTRUGI PRESS
JACKSON, WY

Copyright © 2023 by Vítek Formánek and Eva Csölleová

All rights reserved. No part of this book may be reproduced or transmitted in any form or by any means, electronic or mechanical, including photocopying, recording, or by any information storage and retrieval system without the written permission of the author, except where permitted by law.

Sastrugi Press / Published by arrangement with the authors

Sastrugi Press: PO Box 1297, Jackson, WY 83001
www.sastrugipress.com

Wasted Lives Of Unsung Heroes: Gripping Stories Of How Czechoslovakian Communism Persecuted Their WWII Heroes

The authors have made every effort to accurately recreate conversations, events, and locales from records and memories of them. To maintain anonymity, some names and details such as places of residence, physical characteristics, and occupations have been changed. The activities described in this book are inherently dangerous. The publisher does not have any control over and does not assume any responsibility for author or third-party websites or their content.

NO AI TRAINING: Without in any way limiting the author's [and publisher's] exclusive rights under copyright, any use of this publication to "train" generative artificial intelligence (AI) technologies to generate text is expressly prohibited. The author reserves all rights to license uses of this work for generative AI training and development of machine learning language models.

Any person participating in the activities described in this work is personally responsible for learning the proper techniques and using good judgment. You are responsible for your own actions and decisions. The information contained in this work is subjective and based solely on opinions. No book can advise you to the hazards or anticipate the limitations of any reader. Participation in the described activities can result in severe injury or death. Neither the publisher nor the author assumes any liability for anyone participating in the activities described in this work.

CIP Available
Formánek, Vítek and Csölleová, Eva
Wasted Lives Of Unsung Heroes - 1st U.S. ed.
Summary: Experience the gripping stories of Czechoslovakian pilots who helped to defeat Nazi Germany in World War II only to be persecuted by Communist Czechoslovakia and Stalinist Russia.

ISBN-13: 978-1-64922-325-8 (paperback)

910.4

15 14 13 12 11 10 9 8 7 6 5 4 3

TABLE OF CONTENTS

FOREWORD 7

Chapter 1: **FRANTIŠEK TRUHLÁŘ**
Enemy of Destiny 11

Chapter 2: **JOSEF BRYKS**
Man on the Run 25

Chapter 3: **JOSEF KOUKAL**
Fighter Pilot with Iron Nerves 53

Chapter 4: **GUSTAV KOPAL**
Escaper 71

Chapter 5: **KAREL SCHOŘ**
From Wellington to Communist Concentration Camp 81

Chapter 6: **ALOIS ŠIŠKA**
Six Days in a Dinghy 89

Chapter 7: **ZDENĚK ŠKARVADA**
Keep Floating 108

Chapter 8: **BOHUMÍR FÜRST**
Family Friend 120

Chapter 9: **IVO TONDER**
The Great Escape and Beyond 133

Chapter 10: **BEDŘICH DVOŘÁK**
Forgotten hero 146

Chapter 11: **JOSEF ČAPKA**
One-Eyed Smiling Jo 150

Chapter 12: **ARNOŠT VALENTA**
One Who Didn´t Return Home 164

Chapter 13: **VÁCLAV PROCHÁZKA**
Landing into The Sea 169

Chapter 14: **PAVEL SVOBODA**
Swapped Identity 179

Chapter 15: **OTAKAR ČERNÝ**
Over The Barbed Wire in Two Regimes 184

Chapter 16: **FRANTIŠEK BURDA**
Flying Violin Player in Sagan 192

Chapter 17: **JOSEF HORÁK/JAROSLAV STŘIBRNÝ**
Two Men from Lidice 202

Chapter 18: **JAROSLAV FRIEDL**
From One Hell to Another 209

Chapter 19: **VLADIMÍR SLANSKÝ**
Flight Sergeant "The Broom" 214

Chapter 20: **RICHARD HUSMANN/(FILIP JÁNSKÝ)**
Heavenly Rider 220

Chapter 21: **KAREL ŠEDA**
From Hero to Zero in One Decade 234

LEST WE FORGET 239

SOURCES 243

ABOUT THE AUTHORS 244

ADDITIONAL SASTRUGI PRESS BOOKS 246

FOREWORD

Life should be lived fully and according to the owner's will. Some people do it that way, and some waste it with drugs, alcohol or stay in their jail of choice. But some people had little choice. Those I have in mind were born in the 20s of the last century.

Imagine young 18 years old men from small Czechoslovakia looking into the bright future. The Munich Agreement of 1938 and Hitler's invasion of their country on March 15, 1939, laid waste to their plans. Being educated in strong Masaryk principles and discipline and love for their country, they opted to fight against the Nazis. The only way to do it was to leave the country via Poland to the port of Danzig and then by ship to France or, in a much more complex way, via Slovakia, Hungary, Yugoslavia, Syria, or Palestine to France. They had to sign five-year contracts to stay in the Foreign Legion, where they were trained and treated like scum. When France entered the war, most joined the French Air force, and some fought in the Battle of France. They were the first casualties, and when the French Army quickly collapsed and disintegrated, their only possible way was to get to England, the only free country then.

They were greeted there warmly, and it took some time before Czech squadrons were allowed to be established. However, since

many Czech pilots were highly skilled before the war, they were a welcome addition to the weakened and unprepared RAF in the forthcoming Battle of Britain. They contributed positively; some became highly appreciated aces – e.g., Josef František or Karel Kuttelwascher.

When the 311th Sqn was established, Czechs were joyous because now it was time they could get pay back on Hitler and drop bombs on Germany. The squadron suffered heavy casualties, which still didn't change even after being transferred to Coastal Command. Over 52 airmen were shot down and became POWs in German camps. Many of them tried to escape and fight again but with little success. World War II finished, and those who were fighting abroad could return home. Not in May 1945, but until the August of that year, by which Russian forces had established their presence in the country. Not all families had the luck to see their loved ones back home. More than 530 lost their lives for their country. The survivors were welcomed as heroes, but the following year came the wind of changes, and it got worse in 1947.

After the communist coup d'etat on February 25, 1948, those who fought on the western front were classed as enemies. They were sacked from their jobs and forced to move from Prague. Later, their families were persecuted, kids could not attend universities, and heroes became zeroes. Very often, they were moved into flats of low quality with no running water and inadequate facilities. The only jobs they could do were in factories or hard manual labor. Those higher-ranked were arrested for espionage, sabotage, and allegedly preparing to leave the country, which was made up by the secret police to give them a good reason for levying heavy fines. Most of the airmen were sent to communist concentration or working camps for 3 to 15 years. They suffered from malnutrition, brutal treatment, humiliation, and hard labor. Those destined for liquidation were sent to uranium mines in Jáchymov, where they dug out uranium without any protective gear. Uranium was sent to Russia to "protect the peace in the world." Those arrested were interrogated by Czech Secret Police and brutally tortured to reveal information they didn't know and had no clue about since it

was part of the communist game. Some of them were in cells with former SS men and were humbled by people who spent the entire war at home and now showed sympathy for the leading party and held the "right" political views.

After the death of Stalin and his henchman Czechoslovakian president Klement Gottwald, it took some time before the prison punishments were lowered. In the '60s, most of the prisoners were freed. They enjoyed only a few years of relative freedom before the 1968 Soviet invasion and the arrival of an authoritarian communist leadership, which ended in 1989. By the end of communism, many veterans had died, many were ill, and the youngest was 65. They were morally rehabilitated by president Václav Havel, who was elected. Some got compensation between 2,000-4,500 pounds in today's money. Those who emigrated in 1948 could freely return to their home country after 42 years to meet their friends and relatives. But that was it. They were old pensioners who could only look back and wonder what had happened with their lives and if it was worth it.

They didn't know that our allies, such as Britain, demanded that Czechoslovakia pay in gold for everything Czech soldiers fighting in the RAF or the British Army needed and used. Before the war, the Czechoslovak bank transferred 26 tons of Czechoslovakian gold to British banks. Before this got back to Czechoslovakia, much water ran under the bridge. Only forty-two years later, in February 1982, after years of diplomatic battle, a plane from Zurich finally brought our gold back. However, it was only 18,4 tons. As the former Czech Minister of Foreign affairs, Bohuslav Chňoupek recollected in his memoirs:

"I saw the lists of armament and equipment, and I felt sick. We had to pay for underwear, braces, shoelaces, or handkerchiefs issued to Czech soldiers fighting in England. We had to pay for any bullet or grenade soldiers got. There was even the equipment of those dead killed in the African desert against Rommel, brave paratroopers who assassinated Heydrich, airmen shot down above London and Germany, or those who were transferred from England to Russia."

As former veterans said:

"We were fighting for Britain. But, of course, we were fighting for the freedom of Czechoslovakia. Still, our country was under the German gauntlet, and we were fighting in the English army in England. So we had to pay our Allies with gold to be able to die in the war against Hitler. It was pure business with the blood of dead ones."

So in this book, we would like to tell stories of some brave Czech men who suffered under the Nazis during the war and under the communists in their home country after the war.

LEST WE FORGET

Vítek Formánek, Eva Csölleová

Chapter 1

FRANTIŠEK TRUHLÁŘ
Enemy of Destiny

František Truhlář
In the Czechoslovak army

František Truhlář has been my hero since my school years, and his story has accompanied me for forty years. I studied economics at a school only a few kilometers from Lomnice nad Popelkou, his birthplace, and I befriended his sister Božena. She kept his brother's room tidy, and the bed was prepared with a clean night shirt… in case he had returned…

I could hold his pilot watches which stopped at the very time of his fatal accident in 1946. I could read his personal diary and go through all his pictures. It was THAT that brought me to the topic of Guinea Pigs, which I got hooked on later on. A year after the death of his sister, I received an album full of personal letters and other belongings. There was a rose that he gave to his nurse and love-Pam-fifty years ago. I found her in England with the help of Guinea Pig Club president Tom Gleave. I was hesitant to tell her that I had their war love letters. It was so powerful that I never dared read them.

František Truhlář F.T. (third from right) in East Grinstead

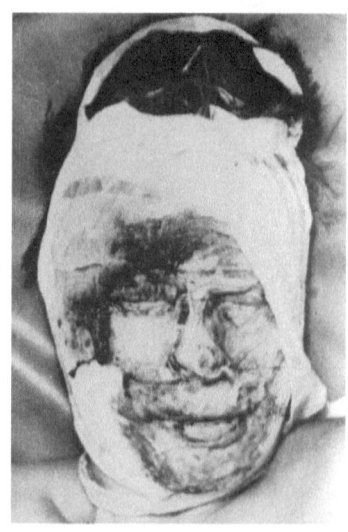

František Truhlář East Grinstead, 13.6.1944

His story is still alive after 80 years. The documentary *Nepřítel osudu* (Enemy of the Destiny) was written and made according to my book and was shown at a film festival in America and won a Silver medal at a festival in Luxembourg. It can be seen on YouTube or www.unitedfilm.cz

František, or Frankie as he was known among RAF mates, was born on November 19, 1917, and had five siblings. His parents wished he would become a teacher. He obliged and graduated in 1936 but couldn't get a teaching job, so he opted to join the army.

He attended the Flying academy in Hranice na Moravě and, in September 1938, left for flying school in Prostějov. But on March 15, 1939, when Hitler occupied Czechoslovakia, the order came for soldiers to return to civilian life. So František returned home but still tried to find the chance to fight against the Germans.

That chance came in January 1940 when he joined forces with Rudolf Turšner (who later joined the armor division) and left home.

Rudolf recollected their escape:

"I met František in Prague through our commander, who knew about our intention to leave. At our D-Day, we met at Prague central station and went by train to Brno and Valašské Meziříčí, where we slept in a hotel and studied the terrain. At 01:00, we got onto a cargo

train and crossed the border into Slovakia. The border was guarded by Slovakian officers accompanied by German patrols with dogs."

We laid under the snow and have seen the guards but managed to get through the Slovakian border, and in Púchov, we jumped out of the train. During the day, we went through the city, and via Trenčín and Nové Mesto nad Váhom, we kept on going to Hungary. We walked only during the night; in the woods, we hid in the deep snow during the day.

That winter of 1940, over six feet of snow fell making it very difficult to get forward. The following night, we were captured by Hungarian guards equipped with machine guns. We were handcuffed and locked in a cellar near the town of Užgorod. The following day, we were interrogated since they thought we were spies.

We were interrogated by the secret police, uniformed police, and soldiers. On the third day, we were escorted to jail in Kosice, Slovakia. For a week, we were locked behind bars, and the Hungarian guards showed us their hostility in every possible way they could. There were five of us in a small cell, and all we were given to eat were peas or lentils mixed with small stones, water, and bread covered with mold.

A trial in Košice decided we had to be escorted to the Slovakian border and returned home. Ten armed soldiers with one officer accompanied us to the border. They motioned us in a specific direction, indicating they would fire if we didn't oblige the orders. Midway through, we all took off in different directions and intended to return to Hungary. Unfortunately, one of us was wounded and spent the whole war in a concentration camp.

František ran behind me, and before dawn, we were back in Košice. There, through ballroom windows, we saw neatly dressed ladies and gentlemen who enjoyed themselves. At the same time, we were soaked, frozen, and hungry. We turned away and walked off.

Out of the blue, two policemen were walking towards us. František stopped and stepped back. I shouted at him, but he didn't listen, so I also turned around and followed him. Police sounded an alarm with their whistles, ran after us, and shot at us with their pistols.

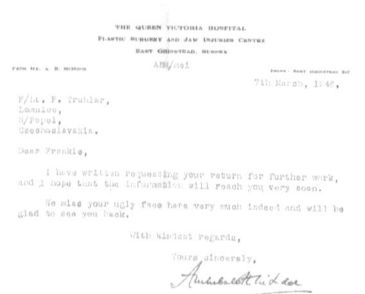

Letter from Archibald McIndoe

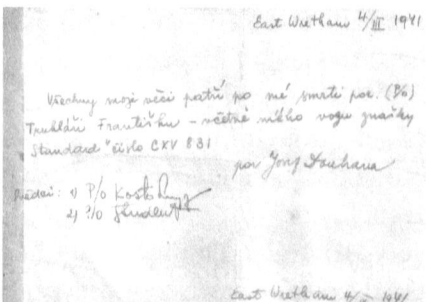

Last will of 311 Sqn friend Josef Doubrava, all belongings will be given to František Truhlář

Thankfully they missed us. We jumped over the fence of a cargo railway station and across piles of coal. We then ran through a field. It was snowing with a strong wind that erased our footsteps. About 3 kilometers from that place, we dug holes into the snow and stayed there until night. When we got up, we walked 30 kilometers on foot that night. František asked me to help him dig a hole in the snow since his knee was severely swollen.

I disagreed with him, so we pressed on and found a barn for about 100 cattle and living quarters next to it. We were both tired and laid under the straw to rest. Our shoes were soaked, so we took them off and waited for about an hour.

A farmer's worker was stamping in his clogs outside the barn, so we got on our feet and waited behind the door. Once he entered, we pushed him into the straw and shut his mouth with his hand. He begged us with his eyes not to kill him, so we freed his mouth.

Frightened, he begged us to leave him alone since he had a wife and children. He milked the cow, gave it to us, and promised to bring us bread. Then he told us his lord was a nice man so that we could rest and dry inside the house. The farmer really was a nice man. He told us to stay here for a few days before we fell asleep. After some days of resting and eating, he took us to the station.

After we got to Budapest, we found the French Embassy. They were interested in qualified pilots who could jump the queue in the transport to France. After six days, I parted ways with František and went solo across Hungary to Romania.

The Yugoslavian border, where I was caught and escorted to a Polish POW camp. I escaped from here, and theatrically, I got into Zagreb in Yugoslavia. I was tired and hungry, and my clothes turned into rags.

With the help of the French Embassy, I got to Beograd (Belgrade). From here, I went with others to Greece, then Turkey then to Beirut, and from here by ship to Tunisia and Algeria to France. I met František again, and he was already donning a French air force uniform."

František arrived in France on March 28, 1940. His dramatic escape lasted more than two months. He was assigned to the air group in Agde and then transferred to Bordeaux. When France collapsed, he flew from Bordeaux to Hendon in England with a few others. He immediately joined the RAF. After the necessary training, he was assigned to the 311th Sqn as a tail gunner.

His story was written by another airman Vojtěch Bozděch in the book Souboj s osudem (Fight with Destiny), published in 1946.

A few more minutes passed, and the port engine of the Wellington bomber had been shot twice and, with irregular droning, shook the whole plane. It was followed by another mighty crack and then another one. At the same time, long sparks started to dance on metal pieces of the plane body. The pale light of fire engulfed the wings' edges. This electric storm was the most feared enemy of all airmen. It waited on the dark horizon while crews thought the danger was already behind them.

Gunner's course 311 Sqn, shortly before the accident., F.T. right

František Truhlář in England

"Oh my God, it starts all over again; what a fuck is going on tonight," said Jan, second pilot into an intercom. The plane was shaking and creeping as if it would disintegrate at any moment.

"We have to go down; otherwise, the port engine will be gone," he said again to Jan, and his voice indicated lots of stress and tension.

The gauges on the dashboard went crazy because of the electric discharges, so it was impossible to figure out the exact height. Not only that, nobody from the crew had the slightest idea where they were right then. Return to base looked rather impossible under current circumstances.

"Skipper to the crew. Guys, it looks bad with us. We have enough juice for about an hour of flight, the gauges went crazy, and we didn't know where we are. We must decide what to do, return and risk unintentionally jumping into enemy territory or staying put, hoping for the best, and trying to get home at any cost. What do you reckon?"

After pondering their chances, all agreed on staying on the plane until the end, whatever that meant.

"Roger, guys. Get ready for anything. I will try to get as close to sea level as possible, and maybe we will have to ditch it. Get on your toes," says Jan into the intercom.

Nobody replied. All understood the pilot's words. At that moment plane leaned towards the left, and a controlled descent started. Jan looked at the gauges, which showed an altitude of 15,000 feet, then 1,000 feet, then zero, and then 5,000 feet again. Both pilots, with Josef in the front turret, looked breathlessly into the darkness under them in the hope that they would break the clouds and they would know if they were flying above sea or land.

Now, like a monster from the horror film, a big wave of seawater opened its mouth as if it was to swallow the whole plane. The next moment, the pilot saw silhouettes of the cliffs before him. He grabbed the plane control and pulled it to his breast to lift the nose of the plane as high as possible.

Now only seconds lay between death and life. The plane jumped over the cliffs, so now they knew they were above land. The big question was, where? The pilots and the crew had to try to recognize anything faintly familiar.

"Guys, we will have to belly land; the tanks are almost dry now; we have no option. Hopefully, lady luck will be on our side. Expect a big impact at any time," says the pilot to the intercom.

Front gunner:

"I don't see any land, but there are some strange silhouettes in front of us."

Pilot replies:

"What do you mean strange silhouettes?"

Front gunner:

"It doesn't look like fighters; it looks as... Fucking hell, balloons. Guys, we are above London."

He shouted in desperation.

The second pilot curses, too, when seeing the monsters ahead.

"We are fucked..."

A bang occurred, a sound of metal being cut…a metal rope cut off the plane's wing, creating a spray of sparks. The plane broke its course and turned upside down–chaos, darkness, cracking, and the whistling sound of the resistant air mixed with desperate screams.

The plane was heading for the ground…a mighty bang…an explosion of dirt, and crashed into a tree. Darkness. Flames.

The demolished plane caught fire, and the wreckage was thrust into an old oak. The broken tail was sticking skywards.

František lay in the burning wreckage of the plane. He heard the sound of flames and exploding ammunition in the turrets. He saw his burnt glove, which was smoking and sizzling on his hand. He moved forward, but something pulled him back. It was his intercom cord still stuck into his helmet. He tore off the helmet, but his damaged face was full of blood which got burnt, and he tore off a piece of his skin, when he took his helmet off.

His lip was cut in two and badly swollen. He tried to find where he was and how to get out. Suddenly he spotted a hole in the fuselage, giving him hope. But the plane's body was red-hot, and every movement was torture for him. He stopped moving due to an immense burning pain in his foot. He discovered his right boot was missing.

Slowly, he crawled through the burning hell, but a kinked strut from the fuselage in front of the hole blocked his path to survival.

František Truhlář on the left

He embraced the brace and tried to bend it to clear the exit.

But the metal strut was red-hot and badly burnt his hands. Totally exhausted, he fell on his back, but the hot fuselage burned his flying jacket, which made him act again.

Slowly but surely, he closed the gap between himself and the hole and got his body from the plane. Since the wreckage was about a meter above the ground, his body weight caused him to leave the aircraft with the fall.

František hit the ground and lay motionlessly for a few seconds. Then he spitted out blood, slowly got on all four, and then, with great pain, stood up on his feet.

Smoke and red heat coming from the wreckage were everywhere. With a groggy walk, he went forward to a small group of civilians who observed the scenery of the tragedy. There was also a priest among them. Then one person turned and in the light of the flames, spotted a burnt silhouette of a man as if it came from hell.

With his scream, he brought the attention of others to it, and all looked in disbelief at the monster coming towards them in slow motion. Finally, the beast collapsed from exhaustion to his knees. The people gathered behind the priest, who came to the burnt airman. It was an extraordinary view, a smoking airman kneeling in front of the priest, who held his hands as if he was blessing him.

That all was underlined with smoke and flames. The kneeling airman pointed his burnt hand towards the wreckage.

"Fi...ve...fri...ends "

Then he collapsed to the priest's feet.

František was transferred to Margareth hospital, where he was tortured with almost unbearable pain from his burns. His bandages had to be changed every three hours, which added to his suffering. Two servants had to hold him when nurse Laura changed

the bandages, and sometimes she almost fainted from the look at his badly burnt and disfigured body.

František was so weakened and tired that he was praying for the end of his suffering once and for all. Finally, he was transferred to a hospital in Halton. The journey was so painful that he fainted. Then came the famous revolutionary plastic surgeon Archibald McIndoe who chose František for East Grinstead hospital, where he tried and experimented with a new form of plastic surgery.

Because of new methods, he tested on his patients - mostly burnt airmen- these guys called themselves the Guinea Pigs and established the Guinea Pig Club. This became legendary since pain was the only way to become a member. In total, 564 patients became members of the world's most exclusive club. František was one of the first members and the first out of four Czechs. (About a dozen members still lived in 2010-V.F.)

During his stay, František saw many badly burnt airmen whose injuries and burns were so bad that he had to have had suffered nightmares from the imagination of what he will see when he was allowed to look in the mirror. At the beginning of 1941, František underwent the first of many plastic surgeries.

The skin grafts from healthy parts of his body were transferred onto the burns. After many months he got the guts to see his new face. With a beating heart, he took a mirror and looked at himself. What he saw made him shake in horror. He couldn't believe it was his face. Scars on the red, burnt face, new deadly-looking lips, and the remains of a deformed burnt nose.

A line of stitches around his eyes kept the eye open more than usual and showed the white of the eye. Eyes separated by deep scars, the bottom part of the face with burnt lower lip deformed the face beyond recognition. He put aside the mirror and thought if this really was reality or a horrible dream. He had only one wish, let it be not true.

But another look in the mirror assured him it's reality. František couldn't imagine life with such a face.

Since Truhlář's burns were so severe, he spent over 19 months in the hospital following convalescence. He met a very kind nurse

Pamela Martin, who he fell in love with, and her friends gave him a massive boost.

She took him downtown among other people to show him that he could live with his burns, and no one cared. McIndoe sent him to London, and František booked a B&B. The lady owner brought him breakfast and told him how her three sons were killed in the war. She refused payment when he asked for the bill and told him how grateful she was that young Czech airmen suffered so much to defend her country.

Author with František's sister Bozena 1991

She hugged František and said to him that after the war, he would be welcomed as a hero in his own country, and she will never forget him. Only if she knew.........

Although no one would blame him if he asked to be discharged from combat duty, he still felt he didn't give Hitler what he deserved. Therefore, František applied to join a lying school to become a pilot. He ended it with an "above average" result and was sent to the Czech 312th Sqn. He flew operationally for nine months and completed 59 combat missions. Then came the 60th, five days after D-Day on June 11, 1944.

On the return trip from France, the weather turned dreadful. They were running out of petrol and were looking for a landing strip. He spotted a piece of land underneath him and informed his number two that he would land immediately. He turned the nose of the plane down, but the engine hiccuped for the last time, consumed the last drop of the fuel, and stopped completely.

The aircraft stopped reacting to the pilot's flying controls and headed toward unavoidable death. The drops of cold sweat appeared on František's burnt forehead.

The short flashback of his first crash four years ago appeared in his mind. The blood was hammering inside his temples, and his eyes opened wide in horror, expecting his forthcoming end.

New skin stripes, transferred onto the burnt face, turned white as chalk.

The plane missed the electric wires by a whisker and hit full speed into the hedge near the street. The impact was so vicious that František saw red circles before his eyes. The plane torn off the hedge, somersaulted, and in many pieces, landed on the very edge of the airfield.

The mighty crack of disintegrating wreckage resounded around. The body of the plane caught fire immediately, and the black smoke was signaling the place of yet another humanitarian disaster. The pilot was engulfed by an infinite depth of darkness and silence. His face, which was so painstakingly made up in East Grinstead hospital, was pouring blood, dripping in rivulets.

Czech pilot would have been burnt to death if it wasn't for the brave acts of two men, Sydney William Thomas and Patrick Mitchell, two members of the local fire brigade. Despite suffering burns, they pulled František's motionless body out of the cockpit and away from the burning wreckage.

František was transferred again to East Grinstead hospital. He was so bad that the priest was called to give him last unction. Nurse Pamela, who got married and quit the hospital service due to an infection, was called to his bedside to help him in his hard time.

McIndoe wasn't sure if his patient would survive the night as he was more on the other side of the river than here. František was having nightmares and utter pain combined with high temperatures. After a few nights of struggle, and with the help of Pamela, who was talking to him, holding his hand, and probably bringing him back to life, he returned to consciousness.

What now worried all concerned people was how to tell him what had happened to him and that he would have to go through the same process all over again. During his long treatment, he befriended his blind mate, a 21 year old airman Jimmy Wright who became very close to him.

When the war was over, František wasn't healthy enough to return to his free country with other mates and was still undergoing further operations. Finally, he could leave England on December

Crash site in Lomnice 1946

1, 1945. Contrary to the reaction of the English public to his scarred face, the response he witnessed in his home country shattered his self-confidence. Once, he returned by train and was alone in his compartment when a mother with a small kid entered. The small baby looked at the face patchwork of the sitting man and started to scream with horror. The mother didn't know what was going on after; finally, her eyes stopped at František's face.

"People like you should not be allowed to go in public, and you only scare people."

František took his luggage and left the compartment with tears in his eyes. He met his RAF mate, Egon Nezbeda, from the 311th Sqn whom he told.

"See, and that will follow me everywhere." From now on, František withdrew and tried to avoid people.

He invited his friend Jimmy Wright to his home in 1946. Jimmy thought traveling to Czechoslovakia in a RAF uniform would be a good idea. He was blind but still wanted to visit his Czech friend. But his uniform was like a red rag in front of a bull's nose for customs officers, and he wasn't allowed into the country. Worse, Jimmy didn't tell anybody about his trip. Archibald McIndoe was surprised that his patient was in Prague, and František asked him by cable for help.

McIndoe replied immediately, that all was fine with Jimmy Wright, and he was finally let into Czechoslovakia. František presented him with glass figures of Snow White and seven dwarfs made in a local glass factory. (When I talked to Jimmy in 1991, he still had them. V.F.) He sent František a parcel with cigarettes, but it came to Lomnice a few days after František fatal accident. When McIndoe read him the cable about this crash, he burst into tears on a hospital bed. He intended to visit his Czech friend's grave again in the mid-90s but sadly died in 1993.

In Truhlář's home, Lomnice and Popelkou, he was a hero for young boys and girls, but some people doubted if he was such a good pilot as he said. Everywhere and every time, people who did nothing during the war criticize people they can't compete with, so at least they hurt them.

It is believed that František once told his doubters he would show them one challenging aerobatic figure. He usually flew his 312th Sqn Spitfire from the airfield in Brno to Prague-Kbely airfield. He flew with former 313th Sqn ace František Fajtl on March 3, 1946, and was believed to tell him.

"František, you go ahead; I will catch you in a minute," and diverted his plane to Lomnice.

It was 13:15, and local inhabitants were having lunch, so the streets were deserted. His plane flew above the village at high speed. František intended to roll the aircraft around its axis, but with his wing, he hit a roof and smashed the Spitfire not far from this house. People, at first, didn't know who it was. Still, his sister Božena found a piece of orthopedic fittings for her brother's shoes, so there was no doubt who the victim was. It is believed that he miscalculated this maneuver, and Lady Luck turned her face to him for the third and last time. There were also rumors that he

Funeral of František Truhlář

Author by the grave of František Truhlář in Lomnice

committed suicide for being looked at as a monster and his Czech girlfriend telling him off.

But František Fajtl denied that saying that František was brave and wanted to live, so suicide wasn't in his cards. In hindsight, maybe it was a blessing for him since, within 15 months, most of his RAF friends were released from their jobs, arrested, and sent to jail. Only God knows how František would end up and how communists would have treated him. He was 29 years old.

I believe he was one of the bravest Czech airmen in WWII.

Chapter 2

JOSEF BRYKS
Man on the Run

Josef Bryks in German POW camp

Looking at the war photos of Czech airmen, I can recognize dozens of faces. However, one I will never confuse with anybody else is the picture of Josef Bryks. His trademark was a deformed officer hat, as if it showed that he never conformed. Josef Bryks and his destiny, a false accusation from the secret police, and a draconian jail sentence symbolize communist oppression towards RAF men and to all who didn't bend their backbones. The way Czechoslovakian officials dealt with his matter is a black mark in our history.

Josef Bryks was born on March 18, 1916, in Lašťany near Olomouc in Moravia. Already at school, he had mastered English which came in handy in the latter stages of his life. He attended the Flying academy in Hranice na Moravě and left it s a Lieutenant/Pilot.

In January 1940, he made his first escape. It was to leave the country and fight against the Nazis. He crossed the border to Hungary and was caught there. After the trial, he was sent to jail in Budapest, where he spent three months in grim conditions and

with terrible food. After release, he continued to Yugoslavia, and in April, he joined the Foreign Legion. It was followed by a long journey across Greece, Turkey, and Syria to the final destination of the French town Agde. After the capitulation of France, he escaped for the second time....... to England. He helped as a translator in the camp for the newly arrived Czech contingent.

After joining the RAF on August 1, 1940 and completing the necessary training, he started his operational career on April 10, 1941, in the Canadian 242nd Sqn. Sadly destiny didn't wish him a long fight against Luftwaffe since he was shot down after only two operational months. On June 17, 1941, he took off from North Weald airfield to escort bombers above France. On the return journey, they were attacked by German ME 109 planes. Josef shot one down, which went down in flames, but he was also shot down and had to leave his burning plane on a parachute. He later recollected that in a protocol signed in London on April 25, 1945:

"I don't remember exactly how I got from the plane. But I felt that the canopy of my parachute opened at 10,000 feet. I felt big pain above my left eye. Burning oil from the plane got onto my face while jumping out. During the descent, I checked the pockets of my uniform to see if I didn't have any papers that could disclose something about my identity. I lost my flying boots either in the plane or while leaving the cockpit, I don't remember that, and my uniform was burnt under the knees. Once I landed, I was approached by French peasants who hid my parachute and gave me a civic coat after finding out my identity. They told me to come at night to a certain place. I couldn't stay there any longer since from the direction of St. Omer came the German patrols on motorbikes. About 300 meters from my landing place, there was a small village where I asked for civilian clothes. At about 23:30, I felt utter pain in my left eye, and I knocked at the house door and asked for a doctor."

Frenchman promised to find a doctor for me, and then he will ponder what to do next. I offered him money if he connected me with some local resistance organization. He sent me to the orchard to wait there. After midnight the German patrol appeared here and found me. They kicked and beat me badly and found my uniform,

CHAPTER 2: JOSEF BRYKS

Josef Bryks first from the left

which I had hidden in the straw earlier. I was taken to St. Omer at about 02:00 and spotted S/Ldr. Smith, from my squadron, and confirmed to Germans that I am a British pilot named Joe Ricks, who was born April 17, 1918, in Cirencester, England.

The following recollection comes from Trudie Bryks, his wife, whom he met three weeks before his last mission and with whom he stayed in touch throughout the war.

After interrogation in Dulag Luft, he was sent to Oflag IV and later to Oflag VI in Wartburg. He wasted no time trying to escape. He had a meat chopper with him with which he dug out a 30 feet long corridor in frozen clay soil and enabled five POWs to escape on the night of April 19, 1942. Josef was teamed with R. Hamilton, and they ran only during the night; they hid and relaxed during the day. Their final and faraway destination was neutral Switzerland. After six nights on the run, a German patrol started to hunt them. The duo hid in the ammunition warehouse. Josef's perfect command of German and workers of all nations helped both escapees. During the air raid at night on 28-29 April, German patrol arrested Hamilton, and Josef was running alone to the closest cargo train station. He hoped to hop on the train and get closer to German borders. Sadly he couldn't get into the carriage, but as he looked around, he spotted parked bicycles. He hid in the shelter, where he stayed for the rest of the night and the following

day. Later that evening, he left the shelter, chose the proper bike, and pedaled away. On the bridge at Geisen, he was stopped by the guard. Still, without stopping, he was zigzagging in the opposite direction and was heading away.

During the short break, he spotted some papers flying in the air, so he grabbed some and figured out they were official invoices on head paper. They probably flew away from some garbage. He put them in his pocket, thinking they may come in handy. When he got to the Offenbach, he was stopped by an armed German patrol. Josef asked to be taken to the commander for him to sign important papers, and he pulled those invoices from his pocket. While the guard went to the commander Josef ran on his bicycle towards the Swiss borders. But the commander set the alarm, and on the next bridge, three armed guards waved to him with stop lights.

Josef didn't stop and pedaled at full speed toward the German trio. He escaped them, but it dawned on him that now they would all be after him, so he went to the forest, hid the bike under the branches, climbed onto a tree, and relaxed. He managed to avoid capture until May 5, 1942, when he started to suffer from dysentery. He was exhausted and found by a Hitlerjunge patrol some 21 miles from Stuttgart. He was sent to Oflag VI b and was supposed to be sentenced for bike theft, but he was so weak and sick that he stayed in the hospital the whole month.

After he was released from the hospital, he spent 15 days in correction. During them, he received razor blades hidden in the bread. When the guard slept at night, he started cutting the wooden door. When he made a small trap door, he began to dig the tunnel, knowing that his cell was at the end of the camp, away from the perimeter, which was controlling prisoners. During the night of 17.8., he made a dummy from his clothes, put it into the bed, and entered the tunnel. Exiting the tunnel on the other end, he was free but only in his pajamas, a woolen sweater, and blanket. He was barefoot and walked in Kassel's direction, hoping he could "borrow" some civilian clothes. The following night, he spotted a plane parked on the small airfield near Frankfurt.

A big dog appeared before him when he climbed the German cockpit guard. Instead of pulling out the gun, he released the dog,

but instead of jumping at Josef, he ran to the guard house for help. Josef climbed from the cockpit and ran away into a small stream to puzzle the dogs. Anti Aircraft battery members caught him during an air raid above Karlsruhe. He was handed over to Gestapo, who brutally beat him. Then he was sent to Oflag VI b Wartburg, but here he realized that RAF POWs were transferred to Oflag XXI b Schubin in Poland. After arriving here, he persuaded commander Harry "Wings" Day to destroy his records so he could start as a new man with a new identity. He became a Polish Jozef Bronisz with a new ID number. Since he could speak Polish fluently, no one could suspect anything. He contacted Polish workers inside the camp with the intent of escaping again and teamed up with Major Morris. They spotted that a Polish worker came in daily with a cart holding a big barrel into which the rubbish was disposed of.

Bryks and Morris obtained gas masks, hid inside the barrel, and waited the whole period before it was packed and waved out of the camp. It was on March 3, 1943, and within an hour, both prisoners were among the Polish resistance. Since Germans were searching for them, "Wings" Day and the other 31 POWs also used that situation and escaped. Most of them were captured soon after. Bryks and Morris waited ten days to get a "green light" from Polish resistance helpers, and they intended to get to Sweden from the Polish port of Danzig. Still, when they heard that is under heavy German control, they opted to go to Warsaw and try to get to Russia. Sadly Morris became ill and waved goodbye to Bryks, who was on his way to Warsaw, which he reached after three weeks.

Meanwhile, Morris was captured by Gestapo, but he persuaded them he was escaping alone. Hence, he diverted sniffing Nazis from his compatriot. Once Bryks arrived in Warsaw, he contacted the local resistance movement. For a while, he worked as a chimney sweeper. But on July 2, the Gestapo raided the place where he was hidden and arrested him and the lady owner of the house. Bryks was handcuffed and taken to the notorious Pawiak prison, where he was told he would be shot as a Russian spy.

He was tortured and beaten but refused to disclose information about his contacts and his collaboration with the resistance. He protested that he was a POW of the RAF but to no avail. Still, the

Josef Bryks in communist jail

day before the planned execution, the order was changed when the Germans in Schubin confirmed his true identity. The lady who gave him shelter, a widow with two small children, was hung on June 17, 1943.

Twelve days later, Josef was escorted to Stalag Luft III in Sagan. On October 13 of the same year, he wrote in his diary, "I was supposed to be repatriated to England as mentally unstable, but three days before that, I was taken to the hospital for an operation on an injury sustained during Gestapo torturing in Pawiak prison."

Back in the camp, he helped organize the Great Escape, which took place on the night of March 24, 1944. (Bryks is said to have escaped from the camp three days later with Australian Wilson, only to be captured immediately after getting out. I was talking about this with Jimmy James-the famous escape artist-who deemed this event as very unlikely, bearing in mind what followed after the Great Escape inside the camp-V.F).

Soon afterward, his true identity was found thanks to the collaboration with his first ex-wife. The Gestapo arrived at the camp and escorted him to Prague. He was confronted by his ex-wife, who confirmed his identity, and Josef was sentenced to death. The execution was to take place on May 2, 1945, in Berlin, but it didn't happen due to the fall of the Third Reich. To wait for his inevitable end, he was sent to the notorious Colditz prison camp and liberated in April 1945 by the American army.

Trudie, the girl he met three weeks before being shot down,

wrote him letters during his captivity, which was a massive psychological boost for him. When the war ended, Josef returned to England and sent Trudie a cable he wanted to meet her. She wasn't sure what to expect since she didn't see him for four years, and not seldom, extended stays in POW camps changed the personality of a human being. Trudie recalled their too-brief and too-painful life in her memoirs:

"When I saw Josef after four long years, he had this wide grin and hugged me. From a young boy became a mature man, but his charm didn't desert him after a long period behind wires. When we walked back home, he asked me to marry him. It was like a bolt from the clear sky, and I was unprepared for that. I couldn't decide until I searched my soul and found out if my feelings toward him stayed the same. Josef agreed, but from the expression in his eyes, I knew he was determined to get me whatever may come. We spent as much time together as possible. After pondering all pros and cons, I agreed with the idea of getting married to him. All the girls envied me since Josef running escapades from the camps became known, and he was to receive Queen Victoria Cross."

Shortly before the wedding, Josef flew home to visit his parents in LaŠťany near Olomouc in Moravia. During the night, Russian soldiers invaded Josef's parent's house in the next village, and people who liked the communists informed them that there was a hidden German pilot. "Liberators" looted everything possible, from horses in the stable to Josef's wedding gifts and his parents' property. Josef was knocked out, and when he regained consciousness, the Ruskies were gone. He alarmed the local police, and they went together to the nearby city of Olomouc. He demanded that the commissar issues him permission to pursue ongoing units. Still, he was drunk and let them wait two hours before he was given permission. Sadly, it was too late since Russians crossed the border with stolen property.

We married in Soho, and I borrowed a wedding dress from Elstree Film Studio. Actress Margareth Lockwood wore the same dress in a movie called "Wicker Lady."

My honeymoon on the Isle of Man started shortly afterward. It was in a hotel in Douglas where I met a fortuneteller. We had a chat and agreed on an appointment. So I took off the wedding ring and went for the session. After a short meditation, the fortuneteller grabbed my palm and said:

"You just married, but he is not an Englishman."

Then she dived into my past and family, and her remarks were spot on. Then she returned to my marriage and told me with sadness in her voice:

"I hate telling you this, but you will be together shortly. I see a very long separation but don't see that you will ever get together again."

I told Josef everything but omitted a part of separation, and as time passed, I let it go off my mind.

Shortly afterward, Josef went to Hammersmith Hospital, where they operated on his war injuries. War ended in the Pacific, so I went to "The Feathers" pub to celebrate it with my workmates. I noticed two men in long black coats who stood by the bar. One of them came to me and invited me for a drink. I politely declined. Then came his friend with an invitation, and I again declined. The first man returned and demanded an explanation why I refused them. It was a silly question, so I told him I was here with my friends

Shortly after, some American soldiers walked directly to our company. They merrily invited us for a drink and celebrated their possible swift return home. The next day, my friend and I visited Josef in the hospital. He looked worried and told me that the previous night he had been woken up by two officers of the Czech secret police who complained about my behavior in the pub and that I was drinking with American soldiers.

We went to the cinema, and the newsreel showed the liberated concentration camps with piles of dead corpses and the motley crew of living skeletons with empty looks in their eyes. It showed subhuman degradation, and even though they survived hell on Earth, they didn't have the power to smile. When the newsreel ended, a well-dressed lady behind them told her partner:

"Don't believe that; Germans are not like that, it is only propaganda, and it was all made up."

I was horrified at her comments and told her sharply:

"Are you crazy? How can you deny such horrible facts? How can you tell it's only made-up propaganda? You should be ashamed of

yourself."

I couldn't believe that after suffering bombings in London and Coventry, there still could be people like this lady who did not feel that Nazism was inhuman.

In October, Josef was ordered to fly to Prague from Blackbush airfield. We kissed goodbye, and I stayed with friends as demobbed WAAF. In the afternoon of that day, I saw the headlines in the papers on the stands.

"Air crash-Czech return-nobody survived."

I was in shock and quickly went through the list of the casualties and thankfully didn't see Josef's name. However, a note about one unidentified body in an RAF uniform who got off the plane at the very last moment. As it turned out, my husband got off on the plane at the very last moment. He was informed by other RAF mates that on board was one Czech lady, a worker for the Czech Red Cross who used its funds for her own benefit and had worn one expensive fur coat while a second one was laid across her forearm. Josef accused that lady and told her he would report this upon his arrival in Prague. He refused to share a plane with her and hopped off at the last minute. It saved his life. Sadly the lady got killed, so she couldn't stand trial, but at least justice was done.

The time to leave England for Czechoslovakia came in October 1946. We crossed the Channel with hundreds of other English women who married Czech soldiers. The journey from Belgium to Germany was stressful when we saw the whole country damaged and thousands of refugees carrying small property on their backs. Finally, after five long days, we arrived in Prague and saw many wives waiting for their Czech partners. Still, sadly there were some "black sheep" didn't come, and as I learned later, one was accused of bigamy.

When we arrived in LaŠťany, our whole family and all our friends welcomed us. I stood at the threshold of new life in a new country with different culture and different language. When Josef announced that I was pregnant, everybody made a toast. I told him my worries since I wasn't ready for maternity and knew little about baby care. He understood me and said my fear was natural

WASTED LIVES OF UNSUNG HEROES

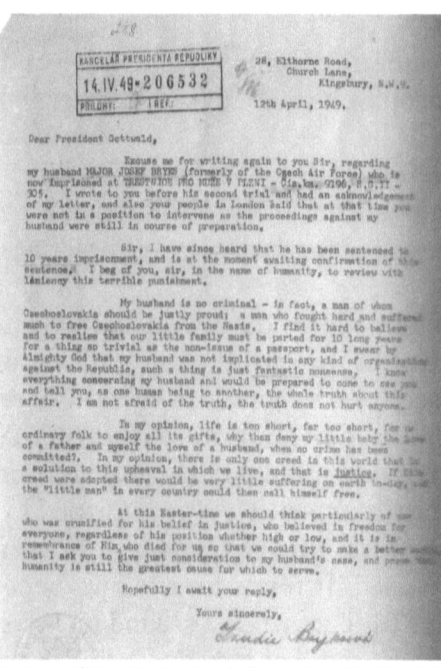

Letter of Trudie to Czech President Gottwald

but would pass shortly.

Nobody spoke English, and I couldn't speak Czech, and although all were very friendly, it was tough for me. They usually spoke German, which I didn't command, and my French didn't help me either.

There was no washing machine, fridge, vacuum cleaner, kitchen sink, soap or detergents, and one big stove where the fire had to be kept all day and night to keep warmth in the room and warm food. The dairy opened at 06:00 for one hour and then at midday for one hour, with no door delivery as I was used to from England. The bakery had similar opening hours; butchers opened only when they had something to sell. Our "Turkey latrine" was crammed outside between stables. I washed in the big bath when everybody was asleep. The tub had to be filled with many buckets of boiling water; this reminded me of scenes from an English film about Welsh coal miners called "How green is my valley."

In December 1946, my sister Dorothy died. I couldn't attend the funeral since Czech authorities didn't grant me a visa. When my daughter informed me she wanted to enter this world, I had to walk to a nearby sanatorium since no cars or taxis existed in Lašťany. Once I got to the entrance, the contractions became urgent. Still, nobody understood English, and I couldn't tell them anything in Czech. But since my state was very critical, formalities went away, and I got into labor and gave birth to a baby girl named Sonia.

Once in the late afternoon, when I went to the bedroom to look after Sonia, I was interrupted by a doorbell. I opened the door, and

there stood two armed Russian soldiers. I knew a Russian garrison was still nearby Olomouc, but I ignored it. Instead, I asked them what they wanted, and they said something in Russian, which I didn't understand, so I told them so in English. Then, out of the blue, he started shouting in German:

"You are not English; you are German; you will come with us."

I started to scream and tried to shut the door, which wasn't possible since one soldier put his boot there. I don't know if my energy or the screaming made the Russians turn back and run away.

This validated the rumors I had heard about the bad behavior of Red Army soldiers during liberation. They were looting, raping, were always drunk, and their primitive habits were disturbing. Some men whose wives and daughters were raped complained to communist commissars. They replied them:

"You should be proud to be able to offer your wives to the brave Red Army."

Knowing that I didn't go to complain since I knew how it would end up and I didn't tell Josef either, for I knew how would he react and what harmful consequences it could bring him.

In May 1947, Josef was to get an MBE from the British embassy in Prague. He was supposed to receive it in Buckingham Palace from The Queen, but the Czech government refused to grant us visas. So I went to England for a short family visit during the summer of that year. Getting the visa was no mean feat, and I got them after a lengthy bureaucratic battle less than 24 hours before departure. After returning to Czechoslovakia, I started teaching English at Olomouc University. I needed and wanted to contribute to the family budget.

Josef got a letter from Colonel Wilson from Australia, who he met in Stalag Luft III and who was offering him a job at his fruit farm business. Josef asked to be released from Czech Air Force, but his request was refused. Later he was told he must not leave the country under any circumstances. We haven't heard more from Wilson, so I assume Josef's letters to Wilson were confiscated and never left Czechoslovakia.

On February 28, 1948, we woke up and heard the shocking news that communists had taken control of Czechoslovakia. Later that afternoon, I returned home from the dentist and saw Josef packing his bag.

"What's up?" I asked him.

"Don't worry, I will probably be arrested; it will be fine," he replied calmly.

He left in the car accompanied by two men in long coats. I didn't know what had happened and wasn't told anything. He was away for more than a week, and still no news. I didn't know what to do or where to call, so my only logical option was to call his superior on the airfield. His answer was cautious and non-committal:

"I don't know anything about him or when he is supposed to be back. Be patient and keep on waiting."

Three days later, I got a cable from Josef, who informed me to come by train to a small village in Central Bohemia, where he would wait for me at the train station. It took me ages to get there. We stayed in a house in the village and in the evening Josef informed me what had happened.

After the coup d'etat, supervised by Russian vice prime minister Valerian Zorin, the new government eliminated all officers who fought on the Western Front who were obviously dangerous to the communist regime. To avoid any possible military resistance from them, they were released from the army and sent to various parts of Czechoslovakia. Since his release, Josef has been cleaning pigsties with Germans waiting for evacuation back to Germany.

He was ordered to attend re-education courses according to the communist line. In his papers, he was classed as unsuited for the political education of his subordinates due to his negative attitude to the success of the communist regime. As he was undesirable to people, he was released for a holiday from March 9 onwards.

After the murder of Czech foreign minister Jan Masaryk, the communist government, without the Czechoslovakian citizens' agreement, closed the borders, built watch towers, and put barbed wire around the borders with Germany and Austria. Passports were void unless approved by the Home Office. In addition, armed guards were patrolling factories to oppress any potential working-class resistance. Later, secret police created fake state borders, many kilometers inside Czechoslovakia, for people who wanted

to escape. Thinking they were already in German, during interrogation, they disclosed contacts to people who helped them, which helped communists to arrest more people opposing the regime.

Josef had to report at the police station weekly and was under permanent control. People - apart from a few loyal ones - started to avoid contact with him and me since it was too risky for them. Once, I spoke with my student who had been absent for awhile, and he whispered to my ear that he had to meet me in one dark street within half an hour. When we met, he looked around, upped his shirt, and showed me his back. They were of fresh scars and blue spots. He explained he was in the demonstration against communist oppression and was beaten up by police. During his release, the prison officials told him that something worse could happen to him if this were to be repeated. He was adamant that he was under control and was spied on.

When I switched on the radio the following day, I listened in horror to the proclamation of a Scottish woman, who believed in a bright future under the communist regime and warned all British women, married in Czechoslovakia, not return to imperialist Britain. Instead, they should stay here and help to build a new, fairer society.

Although Josef and I agreed that I should leave Czechoslovakia first with Sonia and he should follow us because of the worsening political situation, we realized that the opposite order would be better. So Josef left without telling me the details of his escape plan. I think it was for my own safety. Sadly, the escape date wasn't appropriate since the next day was his birthday, and I invited some people to celebrate it with them. It was awkward when guests were in our house, and the main man was missing.

To avoid suspicious questions, I told them he was on the farm and would come later. The company was very merry, but there was a man who I knew only vaguely, and I was suspicious. When 22:00 passed, and he came to me and asked where my husband was, so I explained that he had probably missed the last bus. Guests left before midnight, apart from my inquisitor, who didn't seem to want to leave. He started to make very irritating remarks, so I took every chance to fill his glass with alcohol to distract his

Josef Bryks first from right

attention, finally, at 01:30. I helped him to his feet and showed him the door. Once I got into bed, I heard the knock at the window with a whisper:

"It's me, darling. Let me in, please."

It was Josef. He told me how his first plan didn't come to fruition as his contact didn't obtain concrete-proof documents. The next day there was an announcement on the radio that anybody who won't report a known would-be-escaper would be arrested and sent to jail for a long time. The statement also said that if the escapee were the husband of an English wife, she would be jailed in Czechoslovakia as his ally. Josef took this menace seriously and insisted on me visiting the British embassy and trying to find help for my and Sonia's return to Britain. We went to Prague, and our company was Rhoda and Jo Čapka. Jo was a former RAF pilot of the 311th Squadron and the 68th Squadron, and Rhoda was his English wife, and they also lived in Olomouc, not far from us. Josef was informed that cameras face the entrance to the British Embassy, so he should not be going there.

It was decided that Rhoda and I would go to Embassy while our husbands would wait. Rhoda recognized one embassy employee as a former WAAF officer. She took us to a small office, where we were asked if our husbands intended to leave the country. We carefully answered in the positive that under the condition if they get

their passports they would. She said as much as she knew, namely that they would never get them back. She informed us of a small escape organization that helps former airmen escape to Germany. If our husbands agree, they could meet the organizers at 20:00 at the Swedish Embassy.

We then went there, but two officers only took both Joes to the office while we were left waiting in the foyer. After half an hour, they returned with smiles on their faces. A week after our visit, a signal came for Josef and Jo to return to Prague. They went their separate ways, and I couldn't fall asleep that night. I didn't know then that when they arrived in Prague at the agreed place, they were told the cable should have never been sent since the leak was found in the escaping route, and the attempt was void. So they were advised to return home and wait for another signal a week later.

They again hastily went back home separate ways, and I couldn't fall asleep again. I was nervous but somewhat relieved that no one knocked at my window.

The next day Rhoda came, and we pondered the situation, assuming that if all was well, our husbands could be already in Germany. So, I went to bed peacefully, only to hear a knock at the window after midnight. I wasn't sure if I was already paranoid, so I looked through the curtain and saw Josef's face. I let him in.

He told me he was disappointed that all went well until they were to meet an English woman who was their connection. She came almost 90 minutes late without any excuse. She acted strangely, and general Janoušek, former RAF vice-marshal, who was also in the group, protested against her frivolous behavior. When the German guide arrived, he refused to take six men together under the moonlight, said he would return within four days and left. All men were disappointed with the catastrophic way their escape went and hastily discussed what to do. It was too risky for all of them to stay in the village, so it was agreed they would return to their homes and wait.

Two days later, after 22:00, the two tall men in long coats visited us and demanded to speak to Josef. They were highly positioned Communist officers who went to offer Josef membership in the Communist party and a position in Moscow. They knew

about his stay in POW camps and providing food for starving Russian prisoners. The Party would welcome the man of his caliber, but all were under the condition he would divorce me. Josef strictly refused that offer, and the men warned him that his decision would have dire consequences for him.

I felt the communist loop is getting tighter, and I beg Josef to run away since it is now-or-never. He refused since he felt responsible for Sonia and me, but I persuaded him he should not be afraid. I was sure that with the help of the British Embassy, I would get Sonia and me to England. Finally, he gave up and promised to leave home within 48 hours.

The following evening we sorted out our papers and all the necessary things. In the evening, we went to the cinema for American western. We went to bed at 11 PM. Shortly after midnight, I heard the voices in the anteroom. Josef rushed to the bedroom and told me.

"Darling, it's the police; I am arrested; I must get dressed."

There were three men in long coats, and I could spot they had guns in their pockets.

"Where are you taking my husband? When will he return," I asked them.

"Don't worry, madam, we only want to ask him a few questions in the headquarters," answered one of them.

"What headquarters, where?" I insisted.

"I am sorry, we can't tell you that," answered the same man.

When Josef was ready to leave, he told me,

"You know darling, it's like during the war, at that time though, it was the enemy, now it's our own people."

He hugged and kissed me and put something into my palm. I didn't open it until the police were gone. That paper was American dollars. I ran on the street, and I spotted two black cars. In one of them was Jo Čapka, sitting between two men, and Josef was seated in another car, also between two men.

When both cars left, I stood on the street alone with Rhoda. I made a cup of coffee, and we discussed what to do next and what that all meant.

Early the following day, a secret agent knocked at the door and

demanded to search our house. Even though they had no permission to do this, they had the power, and no one would stand up to them. They could do whatever they wanted, and they knew it. They ransacked everything and didn't care about little Sonia sleeping in the bed. When they insisted on searching my cupboard and personal belongings, I protested they had no right. Without a word, completely ignoring me, they continued with their job. They opened Josef's drawer and pulled out all mail and photographs. All were confiscated, and none of the precious documents or photos have ever been given back to me.

When they were finished, they left, but after about one hour, an officer from the secret police came and started to ask me questions about Josef and also asked about my planned journey back to England, which should happen next week.

"You know more than I do," I replied.

When he realized I wouldn't tell him anything more, he left. As if it wasn't enough, the officer from the Municipal accommodation office came soon after his departure. He demanded the exact date of my departure so he would know when my flat would be free. I told him his information wasn't correct, and I couldn't tell him anything.

Rhoda and I got very stressed, so we went to Olomouc Airfield and demanded to talk to an officer about the arrest of our husbands. Nobody could speak English, so we got nowhere. But we didn't budge and insisted on talking to the officer. It paid off since Colonel Jiří Louda from the local artillery school arrived and spoke good English. He assured us and explained that our husbands were arrested but would be released soon. We told him we wanted to know the place of their stay. He picked up a phone and talked to someone on the other end. After he finished, he said our husbands were still being interrogated. Still, they were treated as officers and will return home within a few days. We only later found out it was only a pile of lies since both Joes never returned to Olomouc, and Colonel Louda got arrested soon after.

The following day, our neighbor told me, we were probably spied on, since a man opposite our house was waiting there for hours. When a postman came and was going to put letters in the box, this man came to him and grabbed the letters. When we realized we got no information from Olomouc officials, Rhoda and I decided to seek help from the British Embassy in Prague. We were told that Englishmen, tied with an escape organization, were back in Britain and won't return here. We were told by embassy officers that they could do nothing for us since it would lead to more significant accusations against our husbands.

When another week passed without any news, Rhoda and I again decided to visit the British Embassy. This time one officer agreed he would go to the Home Office with us. Still, sadly we weren't even allowed to enter the information desk in the lobby. No one wanted to talk to us. Really dejected and annoyed, we spent the whole day going from one police station to another and trying to find any information. Very tired, we ended up at the last station. Officers seemed to understand our broken Czech and looked sympathetic to our situation. The tone of the conversation rapidly changed when a fat man entered the room. He wanted to know why we were there and asked other questions in Czech, which I didn't understand. The policeman behind him gesticulated at me to shut my mouth. When the man left, this policeman explained that he was from the secret police and that every police station had one such man.

During our third visit to Prague, we were lucky since we met one English woman, who knew about our husbands since her husband was arrested with them. They have all been kept under the Prague Castle, infamous jail called "Domeček" (Little House). So, we rushed there and told a prison officer what we were after. He was as stiff as a rock and vehemently denied that our husbands would be there. Then, to put an emphasis on his words, he looked into the book, and with a nod, he said:

"Your husbands are not in this jail."

I didn't believe him and demanded to talk to the director.

"Impossible," was his swift reply.

This answer didn't please me, and I told him I wouldn't leave the place until the director spoke to us. We sat there for a long time and probably became an exotic subject for prison workers who came to observe us. Then the phone rang, and the same officer informed us,

Trudie and Josef Bryks happy for short while

without apologizing, that there was a mistake and our husbands were in this prison.

He said we couldn't see them since it wasn't the official visiting day. Also, the parcels we had with us weren't acceptable. We protested that we had no news for almost a month and that we won't leave until we had a chance to see our husbands.

Since he saw we meant it, he mercifully allowed us can meet our Joe's at 15:00.

We were motioned into a small room with secret agents and a uniformed policeman. They informed us that we were allowed two minutes. I saw my beloved Josef, he was pale and unshaven, and personal conversation was impossible. When a phone ran, and both guards became engaged, I quickly whispered into Josef's ear,

"How do they treat you? What is going on?"

He replied with the same speed, telling me, he was interrogated many hours per day and night. In addition, they switched the lights on and off every five minutes during the night, causing him sleep deprivation and weakening his will to resist.

Two minutes went by far too quickly, and I told him I would come again next week. When Josef has been taken away, officers had told me Josef was accused of many serious crimes.

"What crimes?" I asked in horror.

"Espionage, sabotage, revealing important information to foreign secret services, high treason," he replied.

"These accusations are total nonsense," I protested, before being silenced.

He told me to shut up since the well-being of my husband is in their will and hands. He spoke fluent English, so I asked him:

"You speak impeccable English. Were you in England during the war"?

"Yes, I was; I was in the army," he replied.

"How did they treat you? Well, I guess?"

"Yes, they treated me well, but I would rather forget that time," he replied.

When I stood in the corridor, I looked out from the window. I saw three groups of prisoners in their grey uniforms exercising. Although they looked the same, I was sure I had spotted Josef. I enjoyed looking at him until the prison officer came to me and barked,

"What are you doing here? Don't you know it is forbidden to look out of the window? Get out of here."

We were told that the state had recruited an English-speaking lawyer to represent us. Since the bank accounts of our husbands were frozen after their arrest, we were worried about how we would be able to pay the lawyer. We could spend half in cash and half with meat, vegetable, and eggs. We didn't know then that our husbands would have to pay for their trials. I left the lawyer's office feeling his defense would be a pure formality, and communists would do whatever they liked.

I was allowed another two minutes to talk to Josef. When he kissed me, he told me how badly they were treated and that I should seek help once I got to England. I assured him that things were already underway for me and Sonia to leave the country and I would do my best. Shortly after my return home, I was visited by the secret police. They checked what I had done in the last 48 hours and wanted to know what I talked about with my imprisoned husband.

That evening I invited Rhoda for coffee and talked about our situation. We looked at it from various angles and realized that although we hate leaving our husbands when they need us the most, the problem is getting worse and worse. The longer we stay in Czechoslovakia, the more our husbands will fear for our security. I realized that Sonia would have no future in Czechoslovakia, and we decided we had no option but to return to England.

I visited Josef's relatives and family- since he wasn't allowed to talk to them- to inform them that Sonia and I would return to England after a mutual agreement. When all formalities have been finished, we were told that instead of the regular two minutes, we

would be allowed one hour to say a final goodbye to our husbands. Then we had to visit various offices to tell them when our flat would be free and available for them. We also had permission to take our own property with us. Everything was packed in boxes and parcels.

Our flat was full of customs officers, policemen, secret policemen, and municipal officers who opened and checked everything. I had a strange feeling that this was the very last day I was staying in the flat Josef and I equipped ourselves, and we shared it for a very short time. Josef's father took a horse and the buggy to help us transport our belongings to the train station. We left Olomouc and went to Prague, where we stayed with a friend until departure.

We visited Josef in prison. Sonia, obviously a happy child not knowing what was going on around her, was playing with him, and when I looked at them, I couldn't believe this was for the last time. One hour passed quickly, and the guard entered and strictly informed the prisoners the time was over. Josef kissed me for the last time, and we both cried. He turned back on the stairs and waved me goodbye for the last time.

The day of departure came, and I had all documents issued and my British passport. We boarded the train heading to Ostende, which stopped in Cheb, the last Czechoslovakian town before German borders. Passport control came and took away our passports. They asked us why we travel without our husbands, where are we traveling to, etc. Not satisfied with our replies, they took us off the train and let us wait in a small house near the railways. It took almost an hour of nerve-wracking waiting while officers phoned various places in Prague. Finally, they returned our passports, and we could leave the country.

Forty-eight hours later, we boarded the steamer taking us to Dover. When we met our friends and family, they didn't believe our story, and who could blame them?

Since I had to find a flat for Sonia and me, I also had to find a job. I worked as a secretary in one international company. After about a week, a man with a strange dialect called in and asked for my name. When I asked him what he wanted, he hung up. This repeated every time until I changed a job. Finally, I got a much-awaited cable from Prague that informed me that all of the arrested

Josef Bryks second from left

men, including Josef, were freed. I was happy and started to see the future in brighter colors.

Sadly, it didn't last too long as prosecutor Major Vaš appealed against the verdict, saying that it didn't reflect the crimes they were supposed to have committed- espionage, sabotage, and spying. A new trial was set for February 9, 1949, so eight months later, he insisted that all of the concerned have to be kept in jail.

My husband got ten years of heavy labor in concentration camps. In comparison, Air Marshall Janoušek got 18 years of heavy labor. Also, they were degraded into privates. I was outraged since it broke up all my plans and hope, and I was disgusted by how communists treated honest and pure nationalists, war heroes, and brave men.

In the first month after Josef's imprisonment, his letters came regularly and were not censored. I didn't know where he was kept since the address was some P.O. Box in Prague. After the second trial, the climate changed, and his heavily censored letters came irregularly. Sometimes they were so censored that they didn't make any sense.

My time to do something for him started. I teamed up with Rhoda and other ladies whose husbands were in Czechoslovakian prisons. I visited the Ministry Of Foreign Affairs in London. Although sympathetic to my case, they told me they couldn't do anything since Josef was not a British citizen. They emphasized that

if they did something in my favor, it could even make matters worse for Josef. The Cold War was on. The Iron Curtain was dropped.

I wrote personal letters to President Klement Gottwald, Prime Minister Antonín Zápotocký, Foreign Minister Vladimír Clementis, General Secretary of the Communist Party Rudolf Slánský, Ministry of Justice, and Ministry of Defense.

My plea got to deaf ears, and I received no reply from anybody.

After that, I wrote to Sir Winston Churchill and other important Parliament members. But unfortunately, their answers were in context with the Cold War, and they could do very little. I also wrote to United Nations but got no reply, which disappointed me.

The only member of the parliament who offered me any help at all was William Gallagher from Glasgow. He invited me to his place and told me he was to visit the International Youth Conference in Budapest and could fly to Prague and see what he could do. He knew his mission was risky and couldn't do miracles, but he promised to do his best. We met again after his return home. He told me that communist officers were hostile and told him to "Mind your fucking business."

Mr. Gallagher apologized for being unsuccessful and only prayed his acting won't make matters worse for Josef. Years passed, and I still tried to do anything I could for Josef but had zero positive responses. I also tried to raise young Sonia. A long awaited letter from Josef came in 1954. He didn't write where he was held or how he was doing. He was only worried why I hadn't written him for so long and if there was anything wrong with Sonia.

He didn't know that all letters I sent him during the last three years, were returned to me, unopened. He wrote that his cellmate got a letter from his English wife who demanded a divorce, and he mentally broke down. Josef understood his situation and knew there was no light at the end of the tunnel, so he told me that if I wanted freedom and to lead my own life, he would understand.

"Don't wait for me" were his last honest and desperate words. I was crying above them, and it made me even more determined to get his release, although I knew in my heart that the chances were highly unlikely. I wanted to go back to Czechoslovakia and visit him in jail. The officers of the Foreign Ministry warned me

Author with Trudie Bryks

from doing so since it could be hazardous for me. Moreover, I would have no concrete chance to return since Czechoslovakian authorities don't approve dual citizenship, which I had, and hence I could have been arrested as a Czech citizen.

In 1955, I still wrote letters to Josef, and he came very sporadically and heavily censored. The pauses between them were longer and longer, and in 1956, they stopped arriving completely. Finally, one arrived in 1957, and he wrote me that he was so happy to receive my letter that he almost had a heart attack. Why didn't I write him for so long, did anything happen to me, how is my new job? He never mentioned a heart attack in previous letters, so I took it as a hidden message informing me about his state. I wrote him back but was desperate since I knew only a miracle could help him get out.

In June 1957, I was in the RAF club when the hot news came that Jo Čapka will be released from prison and would return to England. It gave me new hope, and I couldn't believe what I had heard. It was strange that not only was he released from the communist prison, where he sat with Josef, but he was also allowed to go to Britain. I couldn't wait to see him and hear the news from him.

Jo and Rhoda Čapka were our good friends, and I was convinced they would contact me as soon as possible. My friend told me that Jo came to Britain on 30.5. Now they both lived in Essex. June passed, and so did July and August, and I hadn't heard a word from them. At the beginning of September, out of the blue, I got a call. It was Rhoda. I was so pleased to hear her, and she asked me if I had heard from Josef lately.

"No, I didn't. Letters from him came very seldom in past years,

and during the last months, none came at all," I admitted sadly.

"Well, you won't get any more news from him; he is dead," she told me calmly.

I was shocked by the bad news and the cruel way Rhoda had told me. After that, my world disintegrated into pieces.

"How do you know that? How do you know it's true?"

"Well, we have heard that from Jo's family," she replied.

"Oh dear, I am so shocked. I need to meet you and talk to you. I can come at any convenient time for you."

"I am afraid it won't be possible. We are looking for a new house, and we are busy. Gotta go, bye," and she hung up.

I couldn't move and was utterly shattered by the shock of Josef's death and my friend's tone, her cold behavior, and her unwillingness to meet me.

I went to Czech Embassy in London and demanded confirmation of the information and my husband's death certificate. The officer told me he couldn't give me that information and it would take about three weeks to get any information from Prague. I erupted upon hearing that nonsense and shouted at him that if he didn't give me those confirmations within 48 hours, I would contact the British press, most likely The Daily Telegraph.

He calmly picked up two fat files and took page after page. Then, when he went through the lot, he told me:

"From your file, it is evident you are reactionary, and your attitude has helped to imprison your husband. Nothing would have happened if you had been loyal to a people-democratic regime."

He kept reading me the information from my students at Olomouc University, who claimed I was hostile to the Czechoslovakian regime. The whole conversation concluded with his statement:

"There is nothing I can do for you."

The following week I came from work and switched on the TV, where I saw Jo Čapka in the program "This is your life." It was a top-rated program showing many exciting people. Jo was welcomed as a hero who got out of communist jail and managed to meet his wife in England. Sir Archibald McIndoe, a famous surgeon from East Grinstead Hospital, cared about burnt and disfigured soldiers. Since Jo didn't say how he got to England, I assumed he must have

escaped- which was a novelty during Cold War.

Soon after, I visited the RAF club, and the director again informed me that maybe Jo and Rhoda would come today. Everybody was interested to know how he got out of prison. The club was packed, and my friends went in the evening. They took the table in the opposite corner of the club, and I was sure they knew I was there. When the orchestra started to play, Jo came to me and invited me to dance. He cajoled me about how great I looked. I had many questions and asked him what had happened to Josef, what was happening in Czechoslovakia, and how he got out of there.

"I don't know anything about the situation in Czechoslovakia and nothing about Josef. I would prefer if you don't ask me any more questions, for I would prefer not to answer them."

His sharp, negative reply shocked me, and we didn't speak. Jo accompanied me to the table when the dance was finished and went to Rhoda. It was the last time I saw them; our paths never crossed again.

Two months later, I got a letter from Czech Embassy in London informing me that Josef had died from a massive heart attack and was cremated in Jáchymov. Strangely, he died on August 11, while the death certificate was issued on November 6, 1957. A doctor in prison, who did the autopsy, said he never saw such a massive heart attack. The heart was cut in two halves as if done by scalpel blade.

The story of Josef and Trudie Bryks could end up, but it doesn't. Here is part two. While Sonia got married in Germany and later suffered from multiple sclerosis, from which she later died. Trudie moved to Washington, D.C.

After the demise of the Communist regime in 1989 and Václav Havel's speech in Congress in Washington, she wrote him a letter and got a very warm reply.

She decided to visit the Czech Republic and Lašťany after 40 years and find out if anybody from the family was alive. Not only did she find the family, namely uncle Karel Bryks but also friends who knew her husband. Karel told her that when Josef died, a letter from prison informed the family that the body's remains won't be handed over. Instead, the body would be cremated and placed

 CHAPTER 2: JOSEF BRYKS

in an urn with ashes that would be delivered to a family, which never happened.

She returned many times and tied the loose ends with relatives after four decades. She also saw letters that Josef sent his parents from prison, and he supported his old father with the money he earned in jail. However, it was revealed in the 90s that prison officers stole all money he sent.

She tried to accuse former Major Vaš, who sent her husband to jail. But the court didn't find evidence, so Major Vaš wasn't arrested. The irony is that he outlived Trudie.

She also tried to get compensation from the Czech government. It took almost 15 years of legislative tug-of-war before she won the battle. She was a tight fighter like her husband. She received 100,000 crowns which is 3,300 pounds in current money.

Also, in 2009, it was found out that Josef was buried secretly

Author by the memorial in Lašťany

in a cemetery in Praha-Motol, only eight years after his death, so Trudie could lay flowers on the grave of her beloved and brave husband after 52 years.

I befriended Karel Bryks, his family, and Trudie in the late 90s and met them many times. I swapped letters with her until her death on April 28, 2011. She was 91 years old.

There was a wish in her last will that her ashes would be laid in the family grave in Lašťany. So, she, Sonia, and Josef finally got together. Rest in peace.

Chapter 3

JOSEF KOUKAL
Fighter Pilot with Iron Nerves

Josef Koukal - new face

Josef Koukal was born in Jenišovice in Eastern Bohemia on May 6, 1912, as the 8th of 9 siblings. He was very handy, but his family was impoverished. However, he impressed one local businessman who took him under his wing and sponsored his apprenticeship years. Josef wanted to go to the Youth Flying School, so this man persuaded his father to let him go and sign the papers since his son was still underage. There were 600 hundred interested young men, and only 60 were chosen; Josef was one of them. The beginning was tough, and the first thing young would-be pilots had to fill in was a form stating where they wanted to be buried in case of a fatal crash. Not very uplifting for 17 years old young man.

He remembered these days in his memoirs:

"Once, I sat next to the hangar and read a book. I heard a noise and saw my mate flying to the ground in a flat corkscrew. Within a few seconds, the plane smashed to pieces. I put aside my book and ran to the wreckage. My friend was covered in blood, obviously dead,

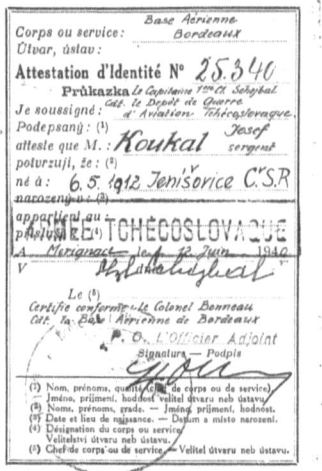 

Josef Koukal - ID card

not a very nice sight. I thought I was done for today, but they called my name within ten minutes."

"I was flying low the other day, and the plane control didn't respond. I was heading into the middle of the hangar. I grabbed the control full force and missed the eaves by a whisker. When I landed instructor came to me and said, "Koukal, go and get drunk; you were born for the second time now."

"These things happened far too often."

He was so good that he became an instructor, and later he was chosen as a test pilot for the aviation company Beneš-Mráz, which produced sports planes. Josef became one of the pioneering aviators in pre-war Czechoslovakia. He married in 1936 a local girl named Františka and almost made her a widow within a few months since he had a few crashes. In 1937, he flew a two-seater prototype of BE550, and after 50 meters, the engine hiccuped for the last time and stopped working. The plane fell head-first to the ground, and the control didn't respond. He kicked the rudder to avoid a strong impact and hit the ground with the wing. The mighty crash followed, and Josef was thrown out of the cockpit and flew 30 meters. Although he had a hole in his neck from a petrol tube, he ran to rescue his mate, who sat unhurt in the middle of a scrap pile. They were taken to the hospital, and the doctor told

Josef that a tube missed an artery by one centimeter.

In 1938, it was apparent that Hitler wanted war, and Josef, as a top ace pilot was on the wanted list of the Luftwaffe. He declined the offer and obviously knew what it would mean if war was declared. However, his wife was pregnant, and he didn't want to put her under stress, so he left. She waited two weeks to give him some time before she went to the police station and reported him missing. It wasn't easy for him to leave in such a situation when his wife would need him, but it was a question of time before the Germans would capture him.

Josef wrote a diary, so his steps and travels are well documented:
I went to catch the train and left Prague on August 13, 1939. We were heading towards Ostrava, and when German guards had a lunch break, we crossed the border to Poland. We were sent to Krakow, where they took my passport and sent me to a nearby camp Bronowice, where I met dozens of other guys. They took us for re-training for Polish planes P.W.S. 2. We returned to Dublin on August 30, where Polish inhabitants welcomed us. We were issued new documents and admired the pleasant local Aviation School. On the morning of September 1, 1939, we were woken up by heavy artillery, which announced the beginning of the war. The following day German Dornier DO17s bombed our airfield at various intervals. I was buried in a trench. It was a nerve-wracking experience, and we immediately lost three aviators and left the airbase.

We crossed the river Visla and slept in the woods. The following morning, nine out of our group told us they were led by a Spanish Civil War officer heading toward the East. We returned to base only to be surprised by the bombing again. The following days were spent in chaos, and the lorry moved us back and forth. Finally, it was announced that the Germans were approaching fast, so we had to move away. On 13.9., seven planes, Potez, and three R.W.D. flew away. It was very foggy, and we couldn't see much, so we landed after a few minutes of flight and stayed in the nearby wood. We met a gamekeeper who told us the Germans were only 13 kilometers from Brest. He took a horse and a cart and told us we could have his lodge, then took some food and his cook and

Home museum in Luze

both ran away. Our Polish pilots set our planes on fire, and we decided to head for the railway station.

We walked through the night and came just in time for the last train to Sarna. Once we got there, we looked for the garrison, which wasn't far. The officer who was distributing weapons invited us for lunch. We were starving, so we accepted his kind gesture, but before we grabbed the bread, over 25 JU-52s flew above us and dropped bombs. Police soldiers hopped into a lorry and left, a Polish guard laid down his gun and left for home, and we stayed there alone. That was the end of the Polish armed resistance.

It wasn't long before hordes of armed Cossacks arrived on their horses and captured us. We had to sit during the rain until early morning in the village green under their machine guns. The following morning a handsome girl looked from the window of a nearby house and asked us:

"Are you Czechs?"

When we confirmed that, she replied:

"One can see civilized men. What are your plans?"

When we said we wanted to fight Germans, she translated that to a Cossacks officer who nodded in understanding. Still, nothing happened. Girl repeated our wish, but the officer didn't reply. Soon after, we were taken to the big bard, where there were other prisoners. The night was very unpleasant, with hunger and stench. Crammed

into the sitting position, we survived the night. A shout and order for the parade woke us up. Opened carriages at the railway station were prepared for us. One Polish civilian told me, "Siberia."

I shouted at the guard that I wanted to speak to his superior, but he ignored me. Then came the officer; I showed him my ID card and told him we were Czech pilots and wanted to join the Red Army to fight against Germans. He took us aside and let us stand there for hours. Then he returned to us and asked what we wanted; I repeated our wish to join Red Army.

He refused, with words, that there was no chance foreigners could join and fight in the Red Army. However, he seemed to be a good man since he issued us papers saying we could return where we lived, gave us a loaf of bread each, shook our hand, and let us go.

We left Sarna and got near the Czech village, Kvasilov, where we searched for a possibility to stay there for a while. We were told it was not possible, so after a night, we continued toward Zborov and then to Tarnopol. The chaos everywhere was overwhelming. We got onto the train and fell asleep, hungry and tired, only to be awakened by a railroad attendant. After telling him our identity, he sent us to a Czech peasant. He let us relax, gave us food, and we told him about our wish to cross Romania's borders. He told his daughter to take me and show me the most accessible border crossing points. She also fixed us a shelter for the night since there was a curfew at 19:00.

The following day, we went past river Dněstr and paid 50 gross to a boatman to take us across the river. Once there, we decided to climb to the forest where we intended to stay the whole day. We met two Russians who grabbed us, took us to a nearby village, and handed us out to men with red tape. They searched us and let the Russian soldiers go. Polish officers escorted us back to Red Army, who locked us in the laundry.

In the evening, a guard brought us soup, wanted to see our ID cards, and told us we had to leave for the 06:00 train the following day. We left, but I left my suitcase in the laundry, so I went back to retrieve it. The suitcase was there but empty; all was stolen, even the shaver. I complained to the Russian guard, who took me to his commander, where I demanded my belongings back, claiming it was a keepsake from my wife.

The officer started shouting at me that there was no stealing in Red Army and that civilians did it. I shrugged my shoulders and was reminded emphatically about the time of my departure. I was teamed up by Václav Kilián (future member of 311th Sqn -V.F.) and one bloke who we didn't know. We walked through the village in a bit of drizzle, and it was just around the 19:00 curfew time.

We sat in the small bush between two bridges that were apart about 300 meters. From there, we observed the guard who walked on the path, stopped, and turned back. It was our moment, so we undressed and naked and got into cold water. After a few steps, a flare shone above us, and the machine guns started barking staccato from the Russian position. Bullets were flying past my ears and slapped into the water. It was a strange moment when I parted with everyone and everything on this Earth. I heard a scream from the left side, where my unknown compatriot was, and I ran away. The flare went off, and the current above my knees was strong. The next flare shone above our heads and the next dose of bullets from Russian machine guns.

I turned around and saw only Václav, on my right, who was monitoring a dark embankment on the Romanian side. I decided to swim in the current and soon felt stones under my feet. I got on the shore and climbed onto the road, where I met Václav. We got dressed in damp clothes and started climbing the sloping hillside. After about a hundred meters, I heard Václav shout, "Joe." So I descended and found him lying on the ground, claiming he couldn't go any further. I stretched my arm to help him get on his feet. We climbed together to the top of the hill, it started to snow, and sleet continued. Václav was shivering, and we still had about 30 kilometers ahead of us. I took him on my back, descended to the village, and asked for help. We passed the first house, and a barking dog alerted an inhabitant. An old lady invited us in. It was very poorly equipped, and we were offered bread and milk. An old lady walked away and returned with a soldier who took us to the camp. The next day we were sent to the police station, where they deported

us to military prison.

After 12 days of stay there, we were sentenced to be sent back. A Polish advocate defended us against the original verdict, and instead we were sent for a month to a medieval prison. We survived 30 days of cruel hunger; sometimes, we had no food for several days. After that, we were sent again with no food to Bazau. Although we were under Polish control, it gave us hope for escape. A major of the Polish Air Force bought us train tickets to Bucharest, where we met some Poles who recognized us and took us to Polish Consulate. They gave us some money and bought us further tickets.

France - J.K. standing

At our own risk, we got to Balchik by the Black Sea. Finally, we passed the control and got on board the ship. Before that, I sent two letters home to the post box. From Romania, we sailed through the Black Sea to Istanbul, the Turkish capital, continued through the Mediterranean Sea, and stopped in the port of Beirut.

We were sent to camp patrolled by Arabs on December 19, 1939, four miserable months after leaving home. Here we met many Czech fellas who had a similar fate. We boarded the cargo ship with oranges heading to Marseilles. We didn't stay here long and went to Bordeaux. Before we managed to get on the front, France capitulated, so we were on the run again. In the port, it reminded me of the chaos in Poland. A Polish captain took us onto the cargo ship, and everybody sat where he could and was in deep thought.

We heard a siren, and our ship was surpassed by Dutch and Belgian vessels. Then we saw some commotion, diverted a course, and increased the speed. Finally, Captain announced that German submarines had sunk those two vessels, and they called SOS.

THE GUINEA PIG CLUB

THE QUEEN VICTORIA HOSPITAL
EAST GRINSTEAD, SUSSEX

MEMBERSHIP CARD

Name: F/Lt. Koukal   Nov. 11 1944

Membership card

"It was heard by a mighty British destroyer who proudly passed us. With no other dramatic situations, we safely got to Falmouth on May 22, 1940. Later on, we were told that our ship was sunk on the return trip to France."

Josef Koukal finally reached his goal; he got into undefeated Britain, where he could fight the Germans. As a highly experienced pilot, he was quickly re-trained in English Spitfires and Hurricanes. He joined the newly established Czech 310th Sqn on June 12, 1940. The First Czechoslovak squadron based on Czech soon got entangled in the Battle of Britain. It was a fight for survival.

Czechs appreciated the warmth of British inhabitants. After so many months of chaos, they saw peace and order, a big morale boost. Being proud patriots, they wanted to show their British superiors they were worth their faith.

At the airfield entrance, there was a giant scoreboard marked with numbers of squadrons operating from that base. Each division had a number showing how many Germans they had on their tally. New 311th Sqn had a big zero in its line, which made Czech pilots ashamed. Josef went onto training patrol and spotted a dot on the horizon. He flew towards that subject, and when he recognized it was the enemy ME 110, he climbed up and behind the German. He tried to get him into the cross of his indicator and pressed the buttons on his machine guns. Nothing happened…the magazines in

the wings were empty since, on training patrols, the rookies weren't issued with piercing ammunition.

Josef gritted his teeth in anger since he could have had his first enemy, which is why he underwent that miserable journey from his hometown. He could put a first scalp on the scoreboard for his squadron.

But on September 3, 1940, the score was already 10:1, meaning that the 310th Squadron lost one pilot and shot down ten enemies. Josef also contributed to the tally, and he again wrote his recollections into the diary.

"One Dornier stayed in my mind. We were having lunch, and before we finished it, we heard SCRAMBLE. Within a few minutes, we were climbing up in our machines and getting into proper formation. The order was to fly onto SE, and we were searching the space around us, awaiting the intruders, planes, and people we learned to HATE. We were at 18,000 feet, and suddenly, we heard "Bandits in the west" on the radio. I looked in that direction and spotted them. Excitement entered me; it's THEM, fucking Jerries. There were plenty of those long-tailed Dornier bombers; one couldn't mistake them."

"Tally ho!" I shouted into the microphone, and finally, the fight was on. Each of us took control of our plane and tried to position it into the cross hairs of our indicators. The air was filled with white lines of lethal missiles fired by the Germans. I spotted an enemy plane in my cross hairs, gave it a four-second burst, kicked out the rudder, and moved away. I saw black smoke trailing from its engine from the corner of my eye. The enemy plane got out of formation and began to descend. The battle soon became separate dogfights, and I searched for another enemy.

When I spotted another Dornier, I positioned my Hurricane for a frontal attack, the most dangerous type. With the enemy plane in the center of my cross hairs, I fired all eight of my heavy machine guns, and the missiles met at one point. My burst shot off the Dornier's right wing. I kicked the rudder to avoid my enemy, performed a back flip, and got behind the Dornier, spiraling towards the Channel. The crew managed to bail out, and I closed

in on the airmen descending on their parachutes. They appeared horrified, no longer resembling the superhuman nation they once seemed. They were nothing more than barbarians who had no shame in bombing cities like London and Coventry, which had no military targets but only innocent civilians. They had no remorse for shooting at crews who had saved their lives. I glared at their faces through my sight with hatred, but I still couldn't bring myself to press the trigger and kill them.

However, I didn't have much time to think about them, as we were soon attacked by their escort ME 109s. We had to hide in the clouds, where I encountered other Hurricanes from my squadron. We received a course back to base and quickly refueled and reloaded our magazines. During the Battle of Britain, we took off multiple times a day, with my personal record being five. It was incredibly exhausting, but the Luftwaffe knocked on Britain's door, and the RAF was firing on all cylinders.

Then came September 7, 1940, a scramble for the entire squadron. I climbed into my new Hurricane, fresh from the factory assembly line. Our orders were to patrol above London. We climbed directly into the sun and saw streams of our planes heading toward the heart of Britain. I spotted the River Thames among the clouds and received further orders over the intercom. We flew 20,000 feet above with our oxygen valves fully open. In front of me, I saw the sky dotted with AA gun explosions and filled with colored shrapnel and white lines of German fighters escorting heavy bombers in tight formations.

I gritted my teeth, kicked the controls to the left, and we flew alongside each other at a safe distance for a short time. The AA guns fired from below, and I wondered why we weren't attacking. Then, I heard a "Tally ho!" in my headphones, and the AA guns suddenly stopped firing, as if everything had come to a standstill. It was as if I was watching a slow-motion film from the outside. After a second, the chaos resumed. I felt like watching a slow-motion movie from the outside. After a second, the hell broke loose, which is difficult to describe in writing.

I found "my" Heinkel and tried to put him in the middle of the

CHAPTER 3: JOSEF KOUKAL

England

indicators cross. My eight machine guns were spitting fire for about five seconds. It was a direct hit. The bomber exploded, and thousands of small parts flew in all directions. I hid in the cloud, and when I flew through it, I spotted the group of nine ME 109 planes underneath me, put on full throttle, and switched on the boost, which I could use only for four minutes with maximum engine power. Thanks to it, the difference between them and me got smaller and smaller. I had already found my victim and was positioning him on the cross.

The AA shrapnel hit my plane then, and the petrol tank behind the engine exploded. The fire got onto the floor under my feet, controlling the pedals. I was surprised at how calmly I evaluated the whole situation. Brain directed me DON'T PANIC. I had the hood opened and undid the belts by knocking the ripcord. I tried to climb from the cockpit, but the strong wing didn't let me out.

I felt that I flew upwards, and my hands grabbed the controls. Unfortunately, the flames got under the dashboard, and now I was sitting in the fire bath. My mind concentrated on maneuvering the plane into the back flip, so I could freely drop out of the cockpit.

Josef Koukal family with author

Although I used all my force to wrest control and leave the plane, natural energy kept me in the gauntlet. The Hurricane was gaining speed, so I decided to try it. All was in vain I gave up. My hands were losing their strength, and the control stick was returning itself. I opened my eyes and spotted bare, burnt hands up to the forearm and raw, burned knees. The remains of the burned trousers were dangling around my calf. The fire burnt more intensely and made a noise combined with the sound of headlong flight. My eyes were stung by needles. HORRIBLE BLAZE. I breathed the warm air, and my mind flew back home. I saw my wife and parents. I was saying goodbye to them.

The second explosion of the main petrol tank built in the wings followed. After that, the plane disintegrated, and a light of hope lit for me.

I was thrown into open space and instinctively dropped my head to my shoulders, expecting to be hit by flying splinters. I had a feeling that I flew with incredible speed upwards. When the pressure seized, I dared to open my eyes; I blacked out for a few seconds. Calm, rational thinking returned to my head. My brain told me to open the parachute, but flames from my burning Mae West and uniform licked my shoulder. "Don't open the parachute" my discretion told me. I had a free fall and was half expecting to hit the ground. But the speed of my fall extinguished the fire, and I opened my eyes to find my position.

I was resting on my back and turning to the right. I estimated my height to be 700 meters and spotted land under me. "Now is the right time to open the parachute," my brain finally directed me. Now it all had to go pretty quickly, but my burned hands didn't listen to me, and it took many precious seconds before I pulled the

ring and opened the parachute. Now I was 300 meters, and I was nicely surprised when I found out it was green grass underneath me. I had time to prepare for landing.

Once I felt the ground under my body, I tried to get rid of the harness, but my hands didn't have the strength to hit the rip cord. I tried for the last time and the third time. Luckily, it worked, and I got the harness off. At that very moment, the wind blew and inflated the parachute, which dragged me behind it. It hurt like hell. Finally, the wind stopped blowing, and I disposed of the harness. Now came the time to get on my feet since my hands and legs were burnt to the bones. I bent my knees and leaned on my left wrist. Doing so, I disjoint all my fingers. Black smoke warned me that my Mae West was still smoldering. I intended to roll back and forth on the ground, but looking at my burnt limbs put me off. I calmly tried to undo the Mae West, and to my relief, the laces and buttons fell off, so getting rid of that wasn't a big problem.

On my left wrist were dangling deformed flying watches. I took them off but tore a piece of burnt skin with it and then took off the rest of the flying helmet.

Then I slowly walked about 200 meters distant from the house. I spotted a man with a hunting gun whom I asked for a car. When he spotted me, he put the gun on his shoulder and opened the entrance gate for me. I spotted a bucket of water by the house. I went for the water but realized I couldn't use the water for my burns. The farmer, Mr. Wright, looked at his wife, who stood by the door to the house and cried. I was observing my skin which rendered down on my hands and turned black. My fingers started to bend, which caused me great pain. Finally, the car arrived and stopped by the gate. I went to it, sat inside, and put my hands on my lap.

The private sat behind the steering wheel, and the officer sat beside him. After about 400 meters, the officer pointed at the debris of my Hurricane. The engine was deeply dug into the ground, and a few bent rods were all that got left. Finally, we came to the village. The officer helped me out of the car and motioned me to the room where was a second officer; he came to me and asked:

"Messerschmitt?"

"No, Hurricane, 310th Sqn Duxford, I am Czech," I replied.

He nodded in understanding and put a morphine injection into my shoulder. Then he took the scissors and cut what was left from my uniform, including my shirt. I sat and tried to remove my boots, but there wasn't much left. I was laid on the stretcher and put into the ambulance.

Life in East Grinstead Hospital - what should I say about that? My hands didn't work; my legs didn't run. I could not see, couldn't move my body, and couldn't speak English. Yet, in my memories, I lived at home, walked around paths, and met known faces, which brought peace into my soul. I underwent 22 plastic surgeries on my face, hands, and legs during my stay there. When I asked for the mirror and looked into it for the first time, I must admit I didn't see a human face. It was more like a patchwork; deep scars gave my face a different outlook. It took 28 long months before Dr. Archibald McIndoe shook my hand and waved goodbye. He was a rare, modest, and pleasant human being."

It was in this hospital where Josef Koukal met František Truhlář. Meeting a countryman in such circumstances was a massive psychological boost for both. Both became members of the famous Guinea Pig Club and experienced something they later missed in their home country, respect from other people who looked at them as ordinary people. One who didn't go through such hell probably can't understand the feelings of the disfigured airman, who has worries about how the public will accept him.

McIndoe instructed inhabitants of the village and prepared them for who and what they could face and meet on the streets. All understood it; hence the return of burnt and disfigured men into society went without significant problems. The body can be healed if the soul is also healed; these two components must work together. It was also why McIndoe was so much loved, respected, and admired by his patients. He repaired their body, understood their soul, and mentally prepared them for what would come. He was an invaluable part of their life from now on.

So when the time came, Josef Koukal shook hands with mates, doctors, and nurses and left the hospital after painful two years.

CHAPTER 3: JOSEF KOUKAL

J.K. in the middle of Guinea Pigs

Here are his recollections again.

"So I left the hospital and was told I did my duty and enough for my country, and I don't have to fly if I don't want to. But I wanted to return to flying, so I underwent medical examinations and went to 53. O.T.U., where we learned the war tactics on Spitfires. After a short holiday, I was sent to the Czech 312th Sqn and got back into combat missions, escorts of heavy bombers, and patrols. The situation on the front changed since my early days in the 310th Squadron almost three years ago. The Nazis had started to lose. But soon, I became tired and was told to leave operational flying. I patted the bodywork of my Spitfire and asked him to forgive me and keep on bringing luck to my mates.

I had problems with my eyes, I couldn't correctly close them, and the pain was constant. They thought it came from the teeth, so they extracted two of them, but it wasn't the right decision since the teeth were healthy, and the pain continued.

Nobody knew then, and I found out after the war that two little 2x2 mm and 1x1 mm shrapnel splinters embedded under my eye weren't ordinarily detectable. So I was sent to a base in Lincoln, where I stayed until the war's end. People sang and danced in Lincoln when V-Day came, and buses were decorated with English-Russian-American flags.

I couldn't wait to get back home, so I went to the Czech air inspectorate in London and asked for a short holiday in Czechoslovakia. We flew from Croydon to Brussels and Prague. When we landed there, I was surprised I didn't see joyous inhabitants, but they looked poor after the prolonged war. I rushed to the train station and went from Kolín to Pardubice on the overcrowded train. I looked out the window and admired every piece of my beloved country. It was like my dreams when I lay in the East Grinstead Hospital. I went by bus from Pardubice to my home village, Jenišovice, and Luže. After that, I went home on foot.

After six long years, I met my parents and my wife again, and I was looking for my son, born on January 18, 1940. He was hesitantly hiding behind his mum's skirt since he didn't know who is that man with a burnt face. On the second day, I visited friends and relatives. I was thinking about that moment all those long war years; now it was reality. After a short break, I returned to England. When we came home in August, sadly not all, since 506 airmen were killed in battles, we marched through Prague, where inhabitants greeted us. President Dr. Edward Beneš thanked us for our contribution to beating Fascists. After he ordered "Dismiss," I went home."

Josef became a war invalid. In 1948 after a long bureaucratic struggle, he got permission to get to England for the eye operation during which McIndoe should have taken out those eight year old metal shrapnel splinters. But doctors considered the procedure too risky and couldn't guarantee that the eye wouldn't be affected. After pondering all pros and cons, the operation didn't go ahead. Josef returned home, where the grim life awaited him.

His wife was made invalid due to heart problems, his second son Peter was born, and all Josef Koukal got from the government for his service to the country was a pension of 450 crowns (15 pounds in current money). The family of four in one room, without a W.C. and running water for 14 years. From time to time, the communist committee visited them and considered moving

CHAPTER 3: JOSEF KOUKAL

Author by the memorial in J.K. home place Jenisovice 3.3.2007

them out and making a barber's shop from their flat. Still, they abandoned the idea when they realized water was not inside the house—Czech Secret Police couple of times tried to provoke Josef to no avail.

Of course, he also considered returning to Britain since he was offered to test Meteor planes. Since he was one of nine children, they all would face massive consequences if he had emigrated. So Josef Koukal didn't mix with other people or go outside his house in Luže. He repaired lawn movers, cars, etc., for his friends but never took any money for that.

He never went to the doctor, and when he had teeth problems, he extracted them with a pair of pliers in his workshop.

In 1964 he was visited by his greatest friend from The Guinea Pig Club, Henry Standen. He showed him around. Josef got an invitation to England every year from the Battle of Britain Association and The Guinea Pig Club, but he was never allowed that. He wasn't the one who would beg communists for a favor; he still remembered when freshly made communists "welcomed" him into the office with words. "How do you want to stay in the army with such face and hands?"

However, when he wrote a letter to England and apologized that he won't be allowed to travel there, the British side made such big waves that within a few days, he got permission with a passport and a financial contribution for a new suit. So he could visit the country which opened its arms to him 25 years ago and meet his mates again. It wasn't his last visit, and during the next one, in 1972, he saw not only East Grinstead Hospital but also the farm where he met the widow of Mr. Wright, who witnessed his accident 32 years ago.

As a memento, he received part of the plane control stick and some debris from his plane, which had been excavated since then. It was a touching meeting, and they had much to discuss.

Sadly, Josef Koukal didn't make it back to Britain again since he suddenly died of a heart attack on February 23, 1980. He died poor and forgotten; that was all he got from his country. He escaped to defend his country against Hitler.

Now, there is a plaque commemorating him on his house, and inside that, there is a small private museum where his sons keep the memory of their brave father alive.

Chapter 4

GUSTAV KOPAL
Escaper

Gustav Kopal in RAF

Gustav Kopal was born on April 18, 1920, in the small village of Jeníškovice. He joined the massive "1000 New Pilots for Republic" campaign but didn't have much time to show off his skill since Hitler dismissed the Czech army. So he decided to leave the country without saying goodbye to anyone. It wasn't wise to inform them about his plans since it could get the family into trouble. On July 15, 1939, he arrived in the Polish town of Krakow. Here he got something to eat, and new document photos were made. From here, he was sent to camp Malé Bronowice, where all Czech would-be-airmen were accommodated. As his recollections say:

"It was a former military camp, typical Polish, which means dirty. About 600 of us, a motley crew dressed in various garbs, from uniforms to tramp gear. Accommodation is worse than bad. I slept with the other two guys on two beds covered with one blanket. We got up at 07:00 to get a coffee, and the food was very basic. He who has money doesn't have trouble. We, airmen, were hated

Gustav Kopal top row, 2nd from left, Sagan

by any other groups, and there were quarrels between Czechs and Slovaks. On 25.7, the transport of 190 airmen left for France. The following transport left the next day; there were 500 people. We faced about 1,000 kilometers across Poland to the port of Gdynia, where we boarded the ship *Chrobry*. After three days on the sea, we entered the French port Bordeaux, and from here, we were herded into a big garrison. Here we were given lunch, one loaf of bread for three men, one can of meat, and one big bottle of beer for four men."

Gustav was accepted into the French Air Force and got shot down and hurt. Still in the hospital, he was informed that France had surrendered and had to run to Africa, from where he headed to England.

On July 24, 1940, he was assigned to RAF in the lowest possible rank, AC2. 11 days later, the first and only Czech bombing squadron, the 311th Squadron, was established. After the required course, Gustav was promoted to sergeant, which enabled airmen to avoid forced labor in case of captivity. He became a gunner in crew which consisted of František Cigoš, Petr Uruba, Arnošt Valenta, Karel Křížek, and Jaroslav Partyk. Then came February 6, 1941. Before the take off, navigator Jaroslav Partyk got sick and

was replaced by inexperienced Emil Bušina from the training unit. The drama of that fatal flight was revealed by pilot František Cigoš after the war during interrogation by Czech Air Inspectorate.

"The take was set for 18:00, and we were ordered to return by 21:00. There was thick fog, but we got successfully above the target and dropped the bombs at 18,000 feet. Then I set the return course back to base and handed the control to the second pilot Petr Uruba. After an hour of flight, the radio operator requested Honington base for QMD and got course 229, which was given to the navigator to be transferred onto the right course corrected by the current wind. The navigator who didn't have combat experience and entered the plane physically exhausted after daily duty, was half conscious due to considerable height and wasn't able to fulfill his task. So I decided to ask for QMD every five minutes and, that way, return to home base in Honington."

Meanwhile, we flew according to the given course 229. Soon after, our radio transmitter/receiver was malfunctioning, so I couldn't get the requested position. The radio operator tried to repair it, but in vain so we were lost.

So I set on new course 270 and descended under the clouds to see where we were. Then I spotted a sea, and at the height of 1,000 feet, we continued to the shore, which we spotted after about 90 minutes of flight. We passed the lighthouse, which we hoped to identify according to the letter, but we didn't manage that.

We ran out of petrol, so we were desperate to land somewhere. I assumed we flew north/west from Wash to the shore and continued in the southern direction. The weather got worse and worse, and the situation got desperate. Suddenly we spotted a small field airstrip and switched on our landing lights. Assuming we were above the English shore, we requested by light signal a position of landing line. We got that immediately.

The landing went smoothly, and as I was moving toward the hangar, about 20 fully armed Germans immediately surrounded us. After realizing our error, I opened the flaps and put on the throttle. Still, sadly the Germans anchored the plane with rope to the ground. According to instructions, the radio operator

Gustav Kopal in France 1940

damaged IFF codes and all important documents. We had no option but to get out of the plane and surrender. We were taken to Fleurs in Normandy, where the first interrogation was conducted. We all changed our names and pretended to be French/Canadians since we spoke French. The next day we were escorted by train across Paris to Frankfurt, where the transit camp was Dulag Luft. We were interrogated for ten days and then sent to Stalag Luft 1 in Barth concluded František Cigoš.

In this camp, Gustav Kopal became an active digger and escaper. Here are his recollections:

Barth was located close to the Baltic Sea and was a propaganda camp for Germans. They could show International Red Cross inspections how well-treated the POWs were. Escapes were organized by Escape Committee, run by Englishmen. I was asked - as a Czech - to keep contact with Slavic prisoners. Many escapes were organized, but all failed, so it was decided that tunnels were the most appropriate way to escape.

Tunnels usually started at latrines or barracks, which was physically demanding. Often we started digging with tablespoons, and it went ahead very slowly. If I remember well, we made four tunnels, but they were discovered by Germans, primarily by mistake. Since Germans figured out I was in touch with Poles, Russian, and Yugoslavians, I was moved to Stalag Luft III in Sagan. Since I was well-built and strong, I was a welcome addition to the digging tunnel department. We made three tunnels called Tom, Dick, and Harry; the last one became known as The Great Escape. But before that happened, I was sent back again to Barth, so my hopes to get away were dashed. But looking at the point that my crew mate Arnošt Valenta escaped and was killed by Gestapo, maybe it was a lucky charm to

Gustav Kopal 2nd from right, France 1940

be sent to Barth before that since maybe; I would have been chosen too; who knows.

After my return to Barth, I thought only about escape. I made a deal with an English doctor who claimed I had bronchitis, so he admitted me to the first aid station. It was positioned outside the Vorlager, which was rounded by barbed wire. I obtained the rags of a Russian prisoner since they were going outside the camp for heavy manual labor. So one day, after I put a dummy into my bed, I got dressed in these rags in the first aid room and mixed into a bunch of Russian workers. I picked up a heavy log and carried that to the coal room. I crawled into the piles of coal and waited there the whole day. In the evening, they counted Russians, and nobody was missing since I mixed with them at midday when they worked close to the first aid room.

Around 23:00, I got out of the coal and crawled towards the last rolls of barbed wire, which I cut with the scissors I had with me. I crawled through them and was outside the camp. It was difficult to describe the feelings, which were a mixture of freedom, joy, and fear. I passed the gulf and headed to the forest, knowing I was safe until the morning when the guards discovered the dummy during morning roll call. I vaguely knew the direction I was going. My aim was to get to France and, with resistance help, return to England. I had sewn documents in my trousers identifying me as Marcel

Floux, French nationality, occupation-high wire fitter who was a forced laborer in Germany.

I hit high fences on my journey, which I didn't count on. When I hit the first fence, I just went around it and did the same when hitting the second fence. I didn't know what that could mean, but I knew that every passing around meant diverting the course. Then I decided that next time I won't go around it. I would climb over it. It was very high, but I still managed to get over it. Once I got onto the ground, I cleaned my hands and intended to continue when a shout "Hande höch" stopped me.

I was scared shitless since moments ago, I had that great feeling of freedom, but now I was in the trap. Behind me was a high fence I had just climbed over, and in front of me was an armed German guard. I didn't want to give up so quickly, but I didn't have a revolver with me, and I was scared that his dog could jump at me. Guard switched on the torchlight, he belonged to Wehrmacht, shouted at the dog, pointed the machine gun at me, and I lifted my hands. That was the end of my hope.

They drove me to the guard house, and interrogation started-where I am from, what do I want here? Since the documents for French workers were sewn into my trousers, I couldn't use them, and I could hardly explain why they were sewn inside. So I tried to lie and said I work for the butcher Zeller in Barth. I remembered that name when they took us to the city, and we passed his shop. I said I was sent for cattle to the nearest village and wanted to cut my way short. Germans answered that I had to stay with them and they would verify my story in the morning. But in the morning Gestapo arrived, and I was accused of espionage since the place I got into was an ammunition store.

They called the butcher, and he denied that he would send someone for cattle and didn't know my name, so I was considered a paratrooper. So when the shit hit the fan, I confessed I escaped last night from the camp. I told them my number, and they called there, but nobody was missing, thanks to my dummy in the bed. Finally, I became desperate and told them, "I am the English pilot Gustav Kopal, and the figure is lying in my bed in the first-aid

CHAPTER 4: GUSTAV KOPAL

Gustav Kopal behind wire

room, call there again, please." They discussed what to do, then slapped my face, and after that, they called there. Ultimately, the dummy was found. They came to pick me up and sent me into solitary confinement for 30 days on bread and water.

When I was released after a month on this camp diet, my friends fed me properly and told me they were digging the tunnel and were quite far. I went to look at it; even 30 days in the cell didn't take my appetite for escape. Finally, I entered the tunnel, which began in our hut, and crawled to the end. My compatriots told me they reckoned they were just right under the wire. So we turned and crawled back. At the entrance, there were Germans. They probably went to check how I was accepted in the hut, so they took the "duty pilot" by surprise. I was beaten again and got another 30 days in solitary confinement on bread and water.

Probably they had enough of me in Barth and transferred me to Heidekrug and then to Mühlberg. There were about 60,000 POWs who lived in very bad conditions. Every week new transports of prisoners arrived, and once we used that chance with my 311th Sqn mate Karel Šťastný. We, RAF sergeants, didn't go to work, so we swapped

Gustav Kopal after the war with ex-mates, from left: Snajdr, Zouhar, Cigos, Sťastný, Kopal, Suza

the numbers and clothes with British soldiers. Germans took us for work in coal mines near Krušné quarry. Here we got mining tools and used to go every day for about 2 kilometers to the mine. After one night shift, Karel and I ran away and rolled down the slope.

We hid behind the bushes until our commando passed us and headed toward the Czech borders. Within three nights, we got to Ústí nad Labem (a Czech town in North/West, close to Germany-V.F.) In the woods, we bumped into the house; the gamekeeper stood outside and talked to his wife in Czech. Then he left for the woods, and we approached the house. The woman got scared since we looked very wild; she gave us food and said she had to send the kids to school. We haven't finished our bread, and here comes a policeman with civilians; they came to arrest us. The woman told the children to report us on their way to school. We were handcuffed and taken to Ústí nad Labem where another interrogation started. Germans took us to Dresden and then to Petscha Palace in Prague. Finally, I was taken to a Court Martial in Torgau and sentenced to death. They took me to Hannover, where I was awaiting my execution.

Thankfully liberators came quicker than executioner."

Gustav Kopal was liberated and sent to England on April 23, 1945. At the beginning of the war, his weight was 72 kilos; after the liberation, it was only 45 kilos. Due to rotten food in the camps, he was diagnosed with diabetes. He returned home on August 22, 1945.

On September 1, 1949, he was discharged from the army by communists. He was offered a job in a kiln but couldn't go there due to his bad health. So he went to work in a rope-making factory in Turnov, but the Czech secret police managed to get him sacked soon after. As if that wasn't enough, he was forced out of their flat with his family and sent to Mělník, where he worked in a waterworks building factory. His new flat was a former porter's lodge which handy Gustav re-created into the living quarter. His family consisted of a wife and two small boys. When the younger one got a fever, and Gustav called the ambulance, the doctor replied that communist society doesn't need children of people like him, and he didn't arrive. The boy later died.

When communist officers saw how well he refurbished the deserted porter's lodge, they moved him out again, this time to a derelict and heath-unfit room in Mlazice. His young son Petr Kopal attended the school. When he finished, he was told he had no chance to go to university; he had only two options: a cooperative farm or the mines. Meanwhile, his father suffered a first heart attack, worsening his health. But the doctor's committee stated that he is perfectly healthy and can do manual work. His wife was also physically sick, so the whole family of three was living on 700 crowns a month (25 pounds in current money).

The big help was his mother-in-law, who helped them as best as she could, and Gustav admitted they wouldn't have survived without her. In 1958, the Communist party sent a letter to the director of the Waterworks building factory demanding Gustav's immediate sack. But the company backed him and only shifted him to a different position. In 1963, there was the famous Globke trial in Germany. Alois Šiška, Vilém Bufka, and Gustav Kopal-all, former 311th Squadron members, were invited to Germany as

Gustav Kopal shot down in France 1940

eyewitnesses. This trial brought international attention to former Czech RAF members, who received partial rehabilitation. Gustav Kopal demanded his Prague flat back, and he succeeded. He found himself a job in the waterworks factory and could get together with his former POW mates in one of the Prague pubs. In March 1968, he got a second heart attack, and he succumbed to that. Five months after that, Soviet tanks invaded his country. One can only wonder what other humility would further normalization have brought him. He just wasn't destined for a life of freedom.

Chapter 5

KAREL SCHOŘ
From Wellington to Communist Concentration Camp

Karel Schoř

Karel Schoř was born on October 10, 1914 in the village Moravské Bránice in Moravia. His father was killed on the front in WWI. He wanted to become a car mechanic, but it wasn't to be. So when he became old enough, he applied to join the army and became a pupil at School for youth pilots. It's worth noting that here he met at least a dozen future RAF mates who either became fighter pilots or members of a bomber squadron. He decided to leave when Hitler occupied his country, and his girlfriend promised to wait for him. He escaped to Poland, and on board of luxury ship *Chrobry*, he sailed to France on July 29, 1939. He was transferred to a French air base in Toulouse. He made two operational flights before the French army signed an armistice with the Germans.

On June 18, 1940, Karel Schoř and four of his Czech mates confiscated French bomber Marcel Bloch M.B. 210 BNS and flew away to England.

He recollected that event:

"Soon after take-off, we flew into the clouds, and it took us some 30 minutes before we found our way out. Above us was clear sky, below us the green ground. I set the course to the sea, and once above then, I changed the course to England. After about 1 hour of flying, our gunner shouted that he saw the land. Later on, ashamed, he admitted it was deja-vu. But we crossed the coast and got above British soil. But soon after that, the first engine seized, and we started to descend. I was so excited and busy that I forgot to check the petrol consumption. We didn't know how long we had flown already, and our arguments were finished by the second engine, which seized too. Now we were flying like a glider, desperate to find some landing strip."

The flight took around four hours, and they crash-landed near Torquay. They were taken to an army camp and waited about a week before the Brits confirmed their identity. A month later, he became a member of the RAF and was assigned to the newly established Czech 311 Sqn, equipped with the twin-engine bombers Wellington Vickers. Almost daily, the members of this squadron took off for air raids above hostile German territory and suffered heavy casualties. Also, the crew of Karek Schoř had many "close shaves", and one of them happened on the return journey from bombing Münster on July 5, 1941. Once they approached the target, the light cone caught them, and they had to zig-zag and corkscrew to avoid precise AA fire. On the third attempt, they finally dropped their bomb load and set the homeward course. Above Dutch Alkmaar at 02:28, they were attacked by German night fighter BF 109s, which caused extensive damage to the plane and hit the undercarriage, dinghy, and wing. The tail-end gunner Ladislav Kadlec bravely defended his aircraft.

Although his turret got a direct hit, he kept firing from his Brownings until he probably shot the enemy plane in flames. After that, he signaled to the pilot that he had been injured. When the captain sent a crew to help him, they had a lot to do to open the damaged turret, which was a right mess. There was a mixture of oil and blood on the floor where the young gunner laid in pain.

His leg was broken, he was dragged out and laid on the floor, and he received a morphine injection. They safely made it home, and Ladislav was transferred to the hospital, where his leg was amputated. But he insisted on returning to action, so equipped with an artificial leg, he returned to 311 Sqn, only to be missing from a patrol above Biscay a year later.

On 9 September 9, 1941, Karel Schoř made a final mission of his first tour, which consisted of 200 hours. He was sent to 1429. Czech Operational Training Flight and took part in a 1,000 bomber raid of Essen. Then he was transferred to Coastal Command, where he sat behind the control stick of a Liberator bomber. While air raids above Germany were somewhat shorter and action-packed above hostile territory, missions above the Bay of Biscay were long - sometimes over 10 hours - patrols, searching for German ships and submarines. After completing the second tour of operations, Karel Schoř was sent to the Bahamas as an instructor, where he stayed until the war's end. His final tally was 83 missions lasting 612 hours, so he contributed significantly to the war efforts.

After returning home in August 1945, he became an airline pilot since Czechoslovak Airlines needed experienced and qualified staff. His girlfriend kept her word and waited for him, and they married in November 1946.

In 1948, he was a plane pilot carrying Czech sportsmen to XIV. Olympic Games in London. Among them were his wife's sisters, Eliska and Miloslava Misakovi, promising gymnasts. Their team won the gold medal, but Eliska got it posthumously. The second day after the arrival, she felt sick. She discovered she had polio, but the communist regime didn't send a doctor with the sportsmen. Before all was organized in London, it was too late, and the young Eliška died before the end of the games. Karel Schoř returned only with an urn of the ashes of his sister-in-law.

Since 1949 he had been visited by two women and one man who asked him if he could carry messages and other things abroad. Karel refused, but in March 1950, he was informed that a man who had visited him was shot during a skirmish at the borders. He was an important man in the net, organizing illegal transfers to

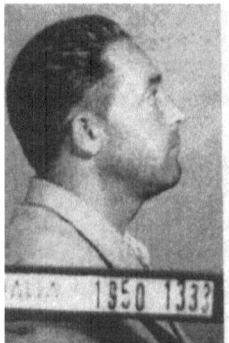

Karel Schoř prisoner in communist camp

Western Germany. It brought Karel Schoř to the radar of the Secret Police. In May of the same year, his passport was confiscated, which meant he could not fly on international lines. On June 27 June 27, 1950, while climbing on board of local flight from Prague to Ostrava, he was arrested by two agents and sent for interrogation. He didn't return home, so his wife was concerned about her husband's well-being. She recollected that:

"Since my husband didn't return home, I called the airfield. I told them I was worried about him. They told me he was sick, which I knew was lying, but I couldn't get any more information and lived in fear. So I went to prison in Prague-Ruzyň to find out if he was there but to no avail. It took about three weeks before one evening at about 23:00 someone knocked at my door. I knew that knock but still asked, "Who is it?" It was my husband; he had lost lots of weight. He told me that the interrogation didn't show evidence of his participation in a criminal organization. The Secret police told him they didn't have any proof against him."

So the next day he went to the airfield to his employer who advised him to take a week's holiday. He was stripped of his flying license and sacked immediately upon his arrival. It was on July 31 July 31, 1950, and three weeks later, he got a job in ČKD Stalingrad in Prague as a turbine fitter."

Since then, he started thinking about how to escape from the communist regime with his family. He intended to get to Austria and then to England. Due to this, he befriended his former

colleague from the 311 Sqn, Gejza Holoda, who was still a pilot of Czech Airlines. Before something could happen, former Czech RAF pilots Jan Kaucký, Eduard Prchal, and Josef Řechka took their families in Dakota to England on September 30, 1950, which led to further actions against former "Westerners." All the rest of them who were still flying in airlines were sacked, and it happened to Gejza Holoda, a potential compatriot in Karel's escape plan.

So in December 1950, a group of former Czech pilots, with their families, decided to leave for Austria with the help of a guide, introduced as Karel Úlehla. No one knew that he was a Secret Police provocateur. They first left on December 14. Vladimír Nedělka was arrested the very day in Břeclav, not knowing that Karel Schoř went there, and it was decided that his wife and others would follow him a few days later. He recollected that:

"I left by train at 18:00 from Prague and reached Břeclav at about 22:00. One hour later, I met a guide in the station restaurant. At 00:30 he got up and took my suitcase, so I followed him. He led me past the railway station towards a river, asking me if I had something for him. I gave him a camera, ladies' watches, and other things, which he accepted."

Then he asked me to wait for him, said he had to find something out, and returned the same way we came. When he returned, he asked me to return with him since the train we intended to hop on would come within half an hour, and it wouldn't be wise to stay at our place for so long. So we went back, and after about 30 steps, we heard the shout.

"Stop." The shooting started when Karel Úlehla dropped my suitcase and started to run away. I laid on the ground and protected my head. It was here where I was arrested."

Not to raise suspicion Karel Úlehla told Gejza Holoda, who was next in line for escape, that Karel Schoř made it abroad and was okay. Meanwhile, Schoř was imprisoned in Prague prison Pankrác. At the same time, his wife was interrogated. She was told, that her husband was arrested and would be charged with espionage and an attempt to escape. The Secret Police also visited her in her flat, which had three rooms. One of the policemen liked it so much that

he forced her to "voluntarily" sign an agreement, that they would swap the living quarters. So she was shifted to the policeman's flat and left within three days. Since she had more furniture than a new accommodation could take, she had it stored away. Still, it was confiscated by police a few months later. Then it dawned on her that she was to be also arrested.

Karel Schoř was found guilty of preparing the escape and potential espionage for foreign intelligence services (communists assumed that he would be asked many questions by foreign interrogators where he would disclose the state's secrets) and was sentenced to 11 years in jail, payment of 2,000 crowns and losing his human rights for ten years plus having all his property and possession confiscated. He was also degraded to Private First Class. An advocate told him:

"You made a good impression at the court. We can't prove anything, but we don't trust you anyway."

The public prosecutor wasn't happy with a low penalty and wanted to revoke and demand a higher sentence.

Karel's wife was also arrested and sent to jail, where she shared a small cell with collaborators and prostitutes. She was found guilty and, as a wife, of a "traitor", who knew about his attempt to escape and not reporting that to the police. She got a six-year penalty.

Other members of the group who wanted to escape were also Battle of Britain pilot Karel Šeda, who was also betrayed by their guide, Karel Úlehla, who was sentenced to 11, 13, and 16 years in jail, respectively. Karel Schoř appealed against his punishment, but it didn't pay off, and the court increased his penalty from 11 to 16 years.

His wife recollected those black years of imprisonment.

"I got to work in a prison workplace near Rakovník. We were accommodated in old stables, where there were 72 of us. We slept on wooden bunks and washed in mangers from which horses drank the water. We did heavy men's labor with a metal press. When we finished the shift, we still had to go to the kitchen to peel potatoes. Once we finished our duties, we were herded to the stables, and doors were locked behind us. We had to pay for our

CHAPTER 5: KAREL SCHOŘ

Karel Schoř with crew

food and accommodation, which was deducted from our payment. I sent what was left to my parents for our son Ivo who was staying with them.

I was freed after 18 months while my husband was still in jail for a long time. My son didn't recognize me and didn't want to leave his grandparents and go with me. I had to fend for myself and our son, so I found a job as a secretary to a social democrat who communists also imprisoned. We lived that way until May 1960, when President Amnesty meant freedom for my husband. So after almost ten years, I could hug him, and we started to live as a family again."

Karel's son Ivo revealed his recollections:

"I was 12 years old when my father returned. The secret policeman used to come to our flat every week to check out what my father did, who he met, and whom he talked to. It lasted for about six months. In the mid-60s, when I wanted to go to university, and although I had good results, I still was not accepted as being a "child of a traitor."

Karel Schoř wanted to clear his name and get his rehabilitation, which partly happened in 1968 due to a political thaw. After that, he could return to flying for Czech Airlines as a second pilot, which he did up until 1974.

After his retirement, he worked as a receptionist at Hotel Budovatel where he could use his language knowledge. He spoke English, French, and some German.

He died in 1986, so he didn't live long enough to witness the fall of communism in his native country. Although he was rehabilitated in 1990 by President Václav Havel, and even Czech Airlines

Karel Schoř in heavy industry after the war

apologized for sacking him in the 50s, it all came too late. It didn't return those ten years wasted in communist concentration camps. Man who risked his life for his country and who spent in the air total of 9816 hours, a man who, among others, got 4x Czech War Cross, 1939-1945 Star, Air Crew Europe Star, Atlantic Star, War Medal, and Defence Medal was only classed as a traitor by his own countrymen in his own country.

Chapter 6

ALOIS ŠIŠKA
Six Days in a Dinghy

Alois Šiška - pilot in 311sq.

Alois Šiška was born on May 15, 1914. His father was killed on the Russian front of WWI on November 18 of the same year. At age four, Alois is said to have announced that he will build a plane and fly to Russia to look for his father. At school, he got a book written by O. Swett Mardel called "Whatever you do, do well" by his teacher, which became his life philosophy.

He built a metal plane model with a precise star-shaped engine as an apprentice, although he hadn't seen a plane in a picture or in reality. His love for flying was evident. Hence, not surprisingly, he opted for Military flying school in Prostějov but wasn't accepted for being underage.

Baťa – a fine businessman and shoe empire boss, selected young Alois for his aviation company. He was working on a new model Zlín XII for which he was offered 5,000 crowns. Still, he chose the pilot's existence instead and became a pilot on September 29, 1936. But before he could blossom into an experienced pilot and use his qualification for the benefit of the mighty Czechoslovakia Air

Force, Hitler made his presence felt. It was obvious that Czechoslovakia would be next in line after Austria's "Anschluss."

Like hundreds of young Czech men, Alois decided to leave his country and actively fight against the Nazi occupation. Since the Gestapo was after him and it was a matter of hours, he pretended to be unwell, got three days of sick leave from the doctor, and told his flat owner he was going to Germany for work. Instead, he hopped on the train to Moravia and got to Hodonín. Alois continues:

"I had a mate Alois Baca with me, and we were lucky since river Morava was frozen, and we crawled onto the other bank quite easily. We went to the pub in the first Slovakian village. We approached the waiter, who understood when we told him the agreed password. He motioned us to the kitchen, warning us that policemen were sniffing around. We got two tickets for the slow train to Bratislava, and for safety reasons, we sat apart. In Bratislava, we bought tickets, exchanged for some pengö (Hungarian money), and boarded the train towards the Hungarian border. We hopped off in Petržalka in Slovakia and went to recommended villa. There were 14 other fugitives. After midnight we left for an essential part of the journey; the wind was biting our faces, everything was frozen, and the snow was up to the knees. The dogs were barking, which we didn't like, but what could we do? Finally, we came to the house where our guide was waiting. There were three times more fugitives than he expected, so he demanded more money. He got one pengö extra from each of us, and we were on our way to the train station. The guide went to buy tickets, and we hid behind a haystack. Guide got back soon but wasn't alone; with him came four policemen.

They herded us to the waiting room at the station, searched us, took all our knives and pistols, and took us to the police station. After interrogation, which lasted about two hours, they escorted us to the border. They took rifles from their shoulders, pointed them at us, and told us to run across the border into Slovakia. Alois and I ran awhile, hid behind trees, and waited. One older man, another fugitive, had a suitcase, so we offered to carry that for him. With

gratitude, he told us he would take us to a person from a nearby village who should help us. We waited until about 04:00 and then crossed the border back to Hungary again.

We got near the train station, but policemen patrolled around the booking office. We opted to hop into the train and buy tickets. To do so, we crawled towards the railway bank in the deep snow for almost one hour and then laid in it for another hour. We got to the train straight from the bank, and for a small tip, the conductor sold us tickets to Budapest. Without further trouble, we got to the French consulate, which was crowded with fugitives. I was given a train ticket to Szeged. I was promised that being a pilot, I should be sent to France very soon, so I got an address where I should wait for the guide to take us to Yugoslavia.

We waited for four days, but our guide didn't turn up. We knew he was supposed to get us a taxi driver with green newspapers, so we decided to find him ourselves. Before we could do so, our taxi driver turned up, and since he showed the required newspaper, we jumped into the taxi and headed to an isolated farm. The farmer was a miserable man who didn't trust us but let us in. He demanded a double bribe. No wonder, since a few days ago, policemen found a group of Czech fugitives on his farm, and he was liable for jail time. We slept on the floor; in the early morning, we got on the sled and went southward.

After two hours ride, we stopped in front of a village. We should pass around it, and behind it was already the Yugoslavian border. One should only go through a small forest. We did so, but we were surrounded by Hungarian border control in front of the border. They took us to their station and then to Szeged jail. Outside temperature dropped well over 20 C below zero. In our cell, there was no window; only through a small hole in the wooden door did a small beam of light get through. One rusty bucket was positioned in the corner, and one bed made from wooden boards was enough for four people.

Every morning they took us for interrogation. Finally, after 11 days of bread and water, they returned all our belongings and called us for a roll call. I was surprised that there were 37 fugitives

who looked very much like me, unshaven and thin. They took us to the old garrison and locked us in the cellar. On the floor were only remains of old straw mattresses. We detested to lay on it and decided to stand the whole night, but we were too weak and tired, so soon, we all gave up.

Our next port of call was called Hodmezövasárhely, and we were locked in an old house. It was a bit better than the previous accommodation, but after three weeks, we figured out we had scabies. We all decided to escape and found out how to do it. We found a small window leading to the yard; we only had to file through the board. But our attempt was halted midway through when one from our group was too noisy; barking dogs alarmed the guards, and that was it. We were transported to the massive stone fortress in Budapest called Citadel.

Soon we discovered about 120 Czechs and the same number of Poles in this prison. Due to the scabies, we had to go to the hospital for treatment, so I decided to try my luck. On the way to the hospital, which was outside the Citadel, I was thinking about where would be the best way to run away. When we passed chemistry, I pointed at my head and the shop to my two guards. I intended to mix with the people and disappear, but my guardian angels were on their toes and turned me in the right direction.

Since it was about midday and the sun was shining, I tried to talk my guards into going for a glass of wine. After a long attempt, we went inside, and I ordered a glass of wine and a second one. When I went to answer the call of the nature-accompanied by one guard - I figured out that the window heads to the yard and then to the street. After a second glass of wine, I ordered cognac for us and pulled out ten pengö, which I gave to a guard and mimed to pay the bill while I went to the toilet again. I climbed through the window onto the yard and the street and jumped into the first taxi.

"Schnell, bitte." The taxi driver understood me but had to start with the handle.

In the mirror, I spotted my guards running from the pub and heading my way. I shouted loudly to the driver for him not to hear guards shouting in my direction. We drove away just in time.

CHAPTER 6: ALOIS ŠIŠKA

I gave him the address of one flat where I once lived for four days, next to the French consulate. I had only five pengö in my pocket, and looking at the taxi meter, the journey cost more. Near the consulate, the driver got out to find the proper address, so I put all my money in the seat and disappeared into the crowd. Then I sneaked by the patrol to get onto the French soil. I felt safe, but not for a long time. The consulate officer told me he couldn't do anything for me and I should come next time. So I asked him if he wanted me to return to the Citadel. He figured out my situation and changed his attitude. He found me in a corner in the corridor behind a lift. I was there alone; six more people accompanied me in the evening. They said two nervous guards and one angry taxi driver were in front of the entrance. It looked to me like a scene from a slapstick comedy. The Consulate officer paid the taxi driver what he requested, but the guard didn't get what they required-me. At 21:00, we left the consulate by the back door and hopped into the prepared car. There was a driver and a guide. I didn't like him, and after previous experiences, I demanded to know the exact plan to cross the borders.

It paid off since, during refilling, the guide left. We drove for a long time, and it was about 02:00 when we arrived at the destination. The driver dropped us off and left, so now we were alone. We had to pass the Hungarian village to the bridge behind, which was already the Yugoslavian border, aim I was supposed to reach some three months ago. Once we left the bridge, I heard the steps "Who is there?" in a Slavic language. We were safe at last.

After two days, we were released, and I could continue to Belgrade, where I went to the Czech national house. I was interrogated and sent to a doctor who would relieve me of the scab. I felt like a human being after a long time."

Alois Šiška took an oath and joined Czechoslovak Army on April 6, 1940, and with the transport of others, went via Greece to Turkey. Here he boarded a small steamship heading toward Lebanon. There he and others from the vehicle were accommodated in the garrison of the Foreign Legion. They went to the sea in the evening and visited a local pub. It was full of drunken soldiers

Alois Šiška - pilot in 311 Sqn

from Senegal who started to touch Czech since they were all gays, and Czechs were later told to be happy to get out unharmed. On April 17, they boarded the transatlantic ship *Patria* which took 160 Czechs, among other soldiers, to the French port of Marseilles.

Alois stayed in France for six weeks, being transferred from one place to another. When France capitulated, his only direction was England. The Czech contingent arrived at Falmouth port on June 24, 1940. They were surprised at how peaceful and organized everything was, unlike in France, where all was in chaos. To show the zero knowledge of the English language of newcomers is the story from a train station where some carriages had the label NO SMOKING, and some had the label SMOKING. Most Czechs went to former ones since they thought the tuxedo was required, not knowing that it divided fag addicts from others. On their journey through the picturesque British countryside, the only thing that reminded them of the war was the painted signposts.

Czech 311th Sqn was established on August 2, 1940, based in Honington, and became fully operational on September 10, 1940. It was a big honor for Czechs who had taken a long trek from their

homes and could fulfill their dream of directly retaliating against the Nazis.

For the rest of 1940 and the whole of 1941, Alois Šiška and his crew flew above Nuremberg, Emden, Hamburg, Berlin, Paris, Kiel, Cologne, Essen, and other targets. One of them had quite an unusual ending. It happened on September 7, 1941, and the target was the heart of Nazi Germany-Berlin:

"We got above Berlin, and it was like a giant firework…. Searchlights, night fighters, AA machine guns…splendid show. It got hot as hell, although outside, it was minus 30 C, and the central heating didn't work again. I looked up into the sky and spotted how searchlights from various angles met at one point, and at the same time, grenades exploded around them. Poor crew; they probably didn't get out of the trap alive. We were at 12,000 feet (3,000 m) and had oxygen masks. Finally, we got to the well-lit target. The navigator asked for a slight change of course. Grenades exploded all around us, and searchlights were looking for their next victim. Wellington jumped up a bit, and I heard in the intercom a refreshing sentence."

"Bombs dropped; let's get out of here."

I took over the plane's control and put engines out of tune to make it harder for German locators. I felt the bomber was lighter and much easier to maneuver. I set the course homeward and felt relieved when flashes and explosions were behind us. Soon afterward, I felt strange tiredness, so I checked the oxygen gauge and discovered the pressure was zero. The rest of the crew also complained about a lack of oxygen, so I sent a navigator to check the oxygen bottles.

"Pipeline is broken by shrapnel, right next to oxygen bottles," was the outcome of his check-up.

Gauges showed a height of 20,000 feet, the sweaty shirt was getting cold, I felt like going to sleep, my lungs were like two pieces of ice, and the hammer was slamming an anvil in my head…A bizarre feeling of getting shrunk.

I set a course 260, put the plane into slow descent…and set up auto-pilot.

I heard nobody in the headphones…darkness.

Then I hit the plane's dashboard with my head, waking me up. My hands were in my lap, and the control moved. I panicked and grabbed it with my hands, but it didn't respond. Then it dawned on me that I switched on the auto-pilot.

"Where are we?"

Nobody answered me. The clock on the dashboard showed 02:00, meaning we flew three hours while we all slept. The crew was slowly waking up. We were at 6,000 feet, and the air already had enough oxygen to breathe without masks.

Navigators asked for a position. We were above the Dutch coast…

For the following two days, we were coughing blood, but on the third day, we were on a plane again for the next mission…such was life in Bomber Command."

Fours days after Christmas 1941, they got the order to bomb Wilhelmshaven. The defense of the port was very heavy, but they managed to drop their bombs and return home. When they left the continent and got above the sea, Alois handed over the plane's control to second pilot Josef Tománek and checked the rest of the crew. Everybody gave him a thumbs-up signal, so it seemed they escaped fate again. But suddenly, the plane was shaken by heavy vibrations, and Alois returned to his pilot chair.

"I checked the engine pressure and saw the gauge jumping down. So I lowered the engine revs and sent Josef to pump the oil. After seeing the first sparks, the engine shattered, and I immediately shut the petrol supply. I called Josef back to the cockpit, since the sparks had turned to flames. The crew doesn't know anything yet, and I think that is good; I don't want any panic. When Josef returned to his seat, he saw the entire scenery. Seconds later, our radio operator shouts into the intercom."

"Fire, we are on fire."

"I know, guys, don't worry; once the petrol from the carburetor burns out, I will extinguish it," I replied.

I pressed the fire extinguisher button the next moment, and the flames disappeared under the white foam.

"Okay, finished; we will have to fly on one engine. Josef is reporting our trouble to Hull" I tried to be as calm as possible.

The engine was switched off, and I heard some rattling sounds. The right engine worked on maximum, but we were still losing height. I couldn't keep the plane in the right direction; we flew in big circles. Navigator was fully occupied, trying to keep a trace of our position on the map. Every minute was precious now, and I tried to keep the plane in the right direction. Another massive vibration shattered the whole plane. I saw in horror how the left propeller broke away from the wing and, with the engine, fell into the sea. I tried to push the throttle, but the control stick returned, meaning the plane was not under control. We were losing speed rapidly.

"Air gunners, leave your turrets. We will have to ditch", I shouted into the intercom, but I got no reply. Instead, I saw the red light on my dashboard, indicating that the front gunner couldn't open the front turret. I sent Josef there, but he misunderstood me, opened the exit door on the floor, and put a parachute on his chest.

"What a fuck are you doing? We are gonna ditch; shut that door."

I shouted at him and wrestled the control stick. We were dropping down at high speed, coming through the clouds. Darkness engulfed us.

"Get ready," I still managed to bark into the intercom and already felt the first touches of the wings in the water.

The plane landed in the sea, and we all managed to get out of the plane, apart from tail gunner Rudolf Skalický. While five of the crew got into the dinghy, the plane slowly sunk into the water, and the rudder was heading skywards like a massive cross on the grave of their unlucky friend. Well, unlucky…if they knew what was ahead of them, they might want to swap places with him.

The dinghy-a yellow rubber boat-was constructed to give the crew a chance to survive a few hours before they are rescued by the Allies or the Axis. It had a signal pistol, paddles, crackers, water, and a first aid kit. But the crew of Wellington KX-B wasn't rescued on either the first or second day. They spent their New Year's Eve.

They spotted a Hudson plane above them which gave them hope, the aircraft saw them and jettisoned a pack that landed about ten meters from the dinghy, but their legs were numb, and they couldn't swim for that. Plane waved at them, which meant he gave a fix to the base, so their hopes raised sky-high. German plane Junkers 88 gave them a burst, a gesture of true barbarism, but luckily didn't hit them.

Thirst, sea-sickness, cold, exhaustion, hunger, and mental desperation took their toll. Navigator Josef Mohr and second pilot Josef Tománek died. Radio operator Josef Šterba was unconscious for most of their stay in the dinghy. Alois Šiška managed to bury the second pilot in the sea, but the dead navigator stayed in the dinghy since no one had enough strength to ditch his body. Finally, only two conscious airmen, Alois Šiška and Pavel Svoboda, the front runner, decided to commit suicide. They mixed the first aid kit pills with seawater and made something, they hoped would be a deadly cocktail and bring them relief.

After six desperately long days and nights in a dinghy, more dead than alive, they were washed up on the shore near the Dutch town of Peten. They were pulled out, given drinking water, and stripped off their trousers. Their legs were frostbitten, numb, swollen, blue-violet, and had septic scars. It was a sad view of the saved survivors. Carefully, they were put into the ambulance and driven to Alkmaar hospital. It was January 4, 1942.

In the hospital, doctors checked them, and although they could not move and were indeed in no state to escape, they were under the control of armed patrol. When Alois Šiška regained consciousness, he asked for water since they didn't drink for six days. When the guard figured out he was an English airman, instead of handing him out the glass of water, he started to strangle him.

The noise alarmed a nurse, who ran in and pushed the armed guard away. The doctors came along with the priest. Since Alois was coming in and out of consciousness and completely exhausted, he fell into a coma and slept non-stop for three days. After some treatment, they all were transferred to a military hospital in Amsterdam. Doctors often came to Alois' bed, and finally, he was put into some tunnel. Only his arms and head were out of it.

His frostbitten feet festered and smelled. The smell was so

Alois Šiška - next to his Wellington

intense that Alois lost his appetite for food. Finally, the German caretaker announced he would rather voluntarily go to the front line than change bandages on his feet. Soon afterward, the feet turned black and doctors decided, they had to be amputated since they caught gangrene.

In the middle of January, he was taken to the operating theatre and got the injection. Somewhat blurred, he saw doctors testing his legs and giving him another injection. When he woke up in his ward, he realized he still had both legs since he was too weak to withstand the operation. He was kept alive only by a hefty evening dose of morphine. The next visitor was the officer from the German intelligence service who demanded some information but got only name, rank, and number. After two weeks, the gangrene stopped and was slowly disappearing. When the head doctor who ordered amputation visited Alois and asked him how he was doing, only to find out he still had both legs, he got furious.

That world is small can be shown in this small story that happened to Alois and his friends in the hospital after Easter 1942:

"One day, they brought into our room an English airman. He was unconscious and was bleeding from his mouth. When he got better, we started to chat. Pavel, our gunner, asked him how they got him."

"We were flying above a German ship but hit its flagpole. The plane exploded and crashed into the Channel. The crew of the ship saved me."

"What type of plane did you have?" I asked.

"Hudson from Coastal Command," replied the airman, who introduced himself as Bill Palmer.

"Don't you know the Hudson squadron with RRA letters?" we asked with poorly hidden excitement.

"That's my squadron," answered the airman without hesitation.

"What were the letters of your plane?" we asked all at once.

"I was a radio operator at RR-P…"

My voice jumped with excitement:

"Fucking hell, you found us in the Northern Sea? Why did you leave us?"

The airman swallowed and started to tell his story:

"I had a week's holiday starting from Sunday 28.12. After my return, I was informed that my crew and two other planes returned from a dinghy search the day before New Year's Eve. Plane RR-P said they spotted a dinghy in the morning hours about 85 miles eastwards from Cromer. He sent the fix to Yarmouth, from where two speedboats set out. Unfortunately, RR-P had to return due to a lack of fuel, and the boats found nothing. So, they were looking for you the whole following day. Then the weather worsened, and the search was canceled."

When the Czech airmen were well enough to be able to withstand a mode of transport, they were sent by train to Frankfurt to the transport camp Dulag Luft.

On October 23, 1942, the Repatriation committee was to decide, who would be repatriated to England. Among those considered patients/prisoners was also Alois Šiška. Apart from German Doctor Jung, there were representatives from Sweden and Switzerland and American and British doctors. When Doctor Jung didn't hear what he wanted to hear from Alois, he turned into an absolute beast:

"Doctor Jung jumped from the table towards me and began tearing off bandages with my skin stuck."

He shouted: "Walk, walk."

I asked the translator to tell the fuming doctor that even he must see that it is not possible to walk on such legs. It enraged him even more. I turned to one of the British doctors to explain the

CHAPTER 6: ALOIS ŠIŠKA

situation with my health. Captain Dickie tried in good German to explain Doctor Jung's situation to me, but the German didn't want to listen. He replied in anger:

"I didn't ask you. You are prisoners in Germany, and I am a boss here."

It was too much for the British doctor, and he informed Doctor Jung that he would complain to British Government about his behavior.

He wanted to add something more but didn't go any further. Jung jumped at the doctor, and he ordered him out. Then he turned onto the guards, who stood behind me and ordered them.

"Pull him backward and forwards. I will teach you to walk, you fucking Czech dog."

They dragged me from the stool to the table and back, and behind me was a trace of blood and pus.

I collapsed. The second British Doctor, Major Charles, wanted to help me, but Doctor Jung didn't hear him, see him, or didn't want to.

He started to shout again, pushed the guards, grabbed me, and tried to drag me again. I was spared the rest thanks to the courage of British caretakers who almost had to kidnap me. Representatives from Switzerland and Sweden informed Doctor Jung that they would report his behavior to their superiors."

In the middle of 1943, Alois was sent to Stalag Luft III in Sagan along with one Canadian airman. At the train station, they were almost lynched by the mob, who realized they were RAF airmen. Armed guards had to protect them against furious German inhabitants.

After only ten days, he was again transferred to the camp at the shore of the Baltic Sea called Barth. He felt here at home since he met another fifteen Czech POWs. But his legs were still in bad shape and festered, so he was sent to a prisoner's hospital near Neubrandenburg. Polish doctors operated on him, and the order for transport came again after a few days. Doctors protested since Alois was not in a condition for a long journey, but Germans didn't have any of that. They only said that the order came from the highest

positions. He was jailed in a cold cell for a month before being transferred to Berlin. As it turned out, the Gestapo finally found him and wanted to bring Czech "traitors" to Prague for court:

"We arrived in Prague on August 22, 1944, and many people were at the station and in front of it. When we were led through them, they whispered in admiration.

"English airmen…"

We wanted to shout, "We are Czechs, real Czechs, and even from Prague."

But we knew what would follow. We ended up not far from the station in Petchka's palace building. It was the HQ for the Gestapo. Our Wehrmacht guards handed us over and left, so our hopes diminished considerably. Luftwaffe treated us with dignity, Wehrmacht respected us, but the Gestapo was different.

Our new hosts in dark black uniforms with death head badge on the hats were intimidating. Their commander entered a room where we stood and shouted, "Face to the wall!" He slapped Eman Novotný, who stood next to me (another 311th Sqn airman V.F.), with such force that he lost balance and fell onto me. I fell onto others, and we were all on the floor in a few seconds like a domino. He grabbed one of us after another one by the collar and pushed us toward the wall. We stayed there hungry and thirsty until the evening without any reason or interrogation.

We were at the end of our power when they grabbed us in the evening. They herded us into the green car and drove us via Prague to Pankrác prison. They wanted us to undress our RAF uniforms and put on old prison rags, which we firmly refused. Our Gestapo guards turned red and wanted forcibly pull off our uniforms. Still, the noise alarmed some big brass, who ran into the room and argued with our guards. In the end, we were allowed to keep our uniforms. We stayed in Prague for over a month, and they didn't get more than a number, name, and rank from us.

In September 1944, we were called up. After so many days in a tiny cellar, we were too knackered that we expected to be executed. But instead, we were loaded into the green lorry, which gave us a spark of hope. It took us to the train station, and we boarded the

train towards Dresden. We got off at a small station and looked in disbelief at the massive old castle with an enormous wall - Colditz fortress. It was the prison for prominent POWs.

Time was ticking slowly, and we felt among British imprisoned officers much safer than in jail. We believed that if Gestapo came to pick us up, it wouldn't stay unanswered and neutral forces would be informed immediately. Every day we stayed alive was a small victory. Christmas passed, yet another war one, then New Year's Eve, the snow melted, and spring came.

One day in April, the German field kitchen came towards the Colditz gate. We were isolated from the rest of the world and lacked food. Our guards couldn't get connected anywhere. We were like a forgotten island in the sea. Suddenly the American Mustang flew past us and made a circle. Then it dived down and sent a burst to the drivers and the guards of the field kitchen. They threw away their riffles and ran away like rabbits. Boxes of cans and biscuits were scattered around. I was looking at that with a couple of my mates and was thinking:

"Well, here, we are hungry. There is food. If we run quickly, we can get the rifles, capture some/the Germans and get the food."

"A couple of us got together and went towards the field kitchen. Once we got closer, we spotted that the Germans were returning. I picked up the rifle but didn't have to point at them. They put their hands up. The food was brought to the main building and distributed among hungry inmates. Soon after, we saw American tanks rolling beneath the castle. They didn't know we were here, so we had to draw attention to ourselves. When the first American tank arrived at the Colditz, and we saw live American soldiers, we believed we survived the war. It took almost two weeks before they could safely take us to Erfurt and from here to Brussels. After that, the war was over; we could return to England."

Alois Šiška got to England and, from Cosford, was sent to East Grinstead Hospital, where Dr. Archibald McIndoe worked on his legs. It was here where he met his badly burnt compatriot František Truhlář. They both were the founding members of the Czech 311th Squadron, and the war parted them for five years, but now

Alois Šiška - early days in RAF

they were both together, marked by war. So, of course, Šiška became the fourth and final Czech member of The Guinea Pig Club. He was still in England in December 1946 when McIndoe came to him and informed him that Frantisek got killed above his house when his Spitfire smashed to pieces. They both were wondering if it was a suicide or just a bad chance.

Alois Šiška returned to his homeland in March 1947, almost 19 months after the others. Of course, the political situation wasn't the same as in May 1945, and he was to find out very soon.

When he reported to the Ministry of Defense, he got a month's holiday to recuperate. Then, after seven long years, he visited Moravia to meet and greet his family and mother. He then had to go to a refresher-flying course in Flying school. One day when he came to his office and opened his locked drawer, he found the form to join the Communist Party. He took it away, tore it to pieces, and threw it into the bin. The same thing happened again the following week with the same result. Then it died out, but the whole town was talking about that.

In July, he was sent to the spa in Slovakia for rehabilitation. While staying there, Secret Police came to him and arrested him. It was in the lodge where he met his future wife, Vlasta. After a few days of investigation and explanation, he was released and could return to the spa. When sacked from the Flying school, he came to the Ministry of Defense to see his chances of getting a job.

Commander told him: "For people like you, there are only three chances: Mines, an Iron-foundry or a Cooperative farm," that was it.

So, he found himself a job in a photo laboratory as a driver. The boss agreed but had to call the personnel department. The clerk,

a loyal communist, shouted that this was not possible and there were only three options for him: Mines, an Iron-Foundry, and a Cooperative farm, and hung up. So, this was a very short lived-job. He had to find anything, so he went to the Minister of Defense, which helped, and was called back to Flying school. Not long after, he was not only forbidden to sit in the plane because of other ex-RAF pilots flying out to Germany, but he was also ordered under house arrest.

Then he got married but was finally sacked from flying school and discharged without severance pay. During Christmas time in 1950, he was forced to leave his flat and move outside Prague. He was sent to a small village with only 27 families. Real shit hole, no bus, no train, no connection, nothing.

So, he obtained old battered German wreckage and turned it into a car. But unfortunately, he could only work on a cooperative farm while his wife could work as an assistant accountant.

Even in such a small village, he was a thorn in the backside of some people. Hot gossips such as Šiška had a wireless receiver or he prints the anti-communist pamphlets and hands them out on the field were made up and spread around. No wonder his house was permanently under the suspicious eyes of the Secret Police. Alois Šiška recollects those dark days:

"To add to that mess was the birth of our daughter. The joy was shadowed by the lack of the fact that we didn't have a washing machine. I built it myself from scrap metal. Once I came home and found my wife in tears. She told me that the Secret Police wanted to confiscate our washing machine, and I was to report the next day to their HQ. They asked me who sold us the washing machine, and when I told them I built it myself, they wanted to know where I got aluminum, which was limited in those years. They didn't believe me when I told them I used it from pistons in my car. I asked them to show me the regulation which would ban me from building a washing machine from scrap metal. They didn't know what to say, so they ended up the discussion with an obvious sentence that summed it up:

"It's up to us to decide what is right and what is not."

Alois Šiška - in wheelchair with Guinea Pig friends in East Grinstead

We worked on the cooperative farm, and they were satisfied with us. But in 1953, the personnel department got a new boss lady. She sacked us immediately when she realized she had two "Westerners" on the payroll. She didn't tell us personally, but in the evening, a car stopped by our house, and the driver told us we were not required anymore and didn't have to go to work in the morning.

My war injuries required further spa treatment, and I have written the request to the local authority. It was turned down by saying that I had no right to that and would only take a place needed by the working class. A new immediate monetary reform was made in May 1953, even though the day before that, president Antonín Zápotocký publicly promised, there wouldn't be any reform at all. People lost all their savings since it was changed at a rate of 1:5 and, in case of more considerable savings, 1:50. Leader of the local communist party changed 300,000 crowns, which showed, how we were all equal, but some were more equal. My new pension was calculated at 938 monthly crowns (30 pounds in current money).

I became a TV repair mechanic just by pure luck. I had a car and was handy, so I was quite in demand. But our luck didn't last long. A communist fruit grower bought our house and moved us out without compensation or alternative living quarters. My wife owned the parents' house in Prague, which a butcher occupied. We demanded her house back, so we would have a place to live. It was turned down, so I went to court. They confirmed my side.

However, the local communist committee wrote that it was not in the public interest that Alois Šiška, a former "Westerner," would move to Prague. In 1960 we found a house in Zvole near Prague, where I continued to work as a TV repairer. Three years later, I

was summoned to the prosecutor's office. They informed me that I was chosen, along with my former 311th squadron mates Gustav Kopal and Vilém Bufka, as witnesses in the trial against Dr. H. Globke, former German state secretary. This trial got considerable publicity, and journalists and TV crews from 23 states were there.

Thanks to this and the political thaw in the mid-60s, I was partially rehabilitated along with other former RAF airmen. In 1965 I was made a Head of Fire, Technical, and Rescue Services on all airfields in Czechoslovakia. I was happy at this job, which gave me a new lease on life. Sadly, after the August occupation in 1968 came a brutal communist gauntlet again, and not surprisingly, I was sacked again in 1970. War injuries and political oppression took their toll, and I had to go to the hospital for further treatment. I realized that I no longer have much power and had to find smaller accommodation in Prague, close to hospitals."

After the Velvet Revolution, Alois Šiška was rehabilitated as all former Czech airmen in RAF. He flew to England many times and welcomed the remaining members of The Guinea Pig Club in the Czech Republic on their regular annual visits. Alois Šiška, along with fighter pilots František Fajtl and František Peřina, became quite celebrities who were invited to the TV, to the press, for autograph sessions or airshows. Initially published in 1967, his book was re-printed in a complete version.

7 Chapter

ZDENĚK ŠKARVADA
Keep Floating

Zdeněk Škarvada - Prague 1946

Zdeněk Škarvada was born on November 8, 1917 in Olešnice. Since his childhood, he has dreamt about becoming a plane constructor. Already at 11 years old, he built his first plane models. At 17, he was allowed - with his parents' blessing - to apply for School for youth pilots in Prostějov. Only 60 boys out of 2500 were chosen, and Zdeněk was one of them. In 1938, he became a pilot and started to work on the airfield in Hradec Králové. On March 31, 1939, all aviators were released from the army, and the Wehrmacht confiscated airports and planes.

On June 8, 1939, he decided to leave Czechoslovakia and fight against Hitler. Via Ostrava, he got to Krakow and camped in Malé Bronowice in Poland, a meeting place for likewise-minded Czechs. Next, he traveled to Gdyně port and was about to board the ship *Castelholm* which would take him directly to France. While waiting in the queue, Polish Air Force officials offered young Czechs a place in their army. Most refused that; some hesitated

and didn't know what to do. Finally, Zdenek picked up a Polish coin and made a toss. When he picked it up, it was decided, Poland. So he left the queue and, with twelve others-among them was also Josef Frantisek, future Battle of Britain ace, boarded a train to Dublin, where was the Central Flying School. All were accepted to Polish Air Force on July 26.

Polish soldiers took the bets that war would start soon, but despite that, there was not a single machine gun to defend the airfield, no shelter, and no trenches. So when the war started on September 1, 1939, it caught Poles totally unprepared. Airmen flew obsolete planes RWD 8 from the airfield to airfield, but everything turned into chaos. Refugees' army supplies were non-existent and there were overcrowded streets.

Zdeněk recalls that time:

"Once, we flew in a group of RWD-8, and I ran out of petrol. I had little time left, so I landed in a potato field. It wasn't an ideal place, and the undercarriage got stuck in the mud, so I thought that for me the war had ended sooner than it had begun. But out of the blue, my friend and mate Josef František spotted me and made a round before landing himself. He shouted and gesticulated at me to run to him and get on the plane. He was sitting in the back seat, and in the front one, he had luggage.

"There is no space for me," I shouted back.

"It doesn't matter; get in somehow," he insisted.

So I climbed up, wedged my legs among the luggage, sat on the cockpit's rim, and held the wings struts. I don't know how he managed that, but Josef put the full throttle and managed to take off. German tanks were not that far, so he definitely saved me. It was my last flight on Polish soil."

Once he was in the relatively safe zone, the plane landed. Zdenek continued with four other mates by foot to Romania's borders. On September 17, 1939, Red Army entered Ukrayina and divided Czechs and Poles into two groups. Zdeněk obtained a new ID card since Russians classed everyone without any documents as potential spies and sent them to camps. He continued with others from Volyň towards Czech Kvasilov, where Czech organs cooperated with

Russians in organizing transport for refugees to France. In March 1940, according to order, the group continued to the center of Russia. The journey by train lasted almost a week and went via Kyjev, Kursk, Voronez to Suzdal. Watch towers and barbed wire around the camp, and inside, number of Czech was about 850.

Zdeněk remembered those days in Russia with a sarcastic chuckle:

"We were all very depressed when we realized we would be herded behind barbed wire and watched by about 120 Russian guards. The only hope for us was that this wasn't either Gulag or a concentration camp but only an internment camp. We were starving, and the promise to be fed was pushed back and forth. We were divided into wooden houses. When the lorry arrived, we were cheering that the food had finally arrived. Instead, we all got one issue each of the Constitution of the USSR and the Story of Good Soldier Svejk, of course, all in Russian. With empty stomachs, we were sent to bed."

The following day, we were all sent for the delousing procedure, which we welcomed. Still, the sadly very primitive method resulted in one batch of clothes being burnt to dust. A group of inmates had to stand bare-naked until a suitable replacement was found. We had to do the chores daily, and the Russians were pretty helpful. Another duty we had to do every day was help in the kitchen, peeling potatoes, scratching fish, etc. After a period of relative calm, Russians forced us to do ordinal training every day and opened the school for COs and NCOs. The unity of our group was dissolved into three groups secretly observed by Russians.

The biggest group was the so-called "French," about 400 airmen who wanted to go to France and start fighting Germans. I was among them. The second group, with about 200 men, was called "Astronomers," which was left-wing orientated and refused to participate in what they called an imperialist war. They wanted to stay in Russia. They wore red stars on their hats. The last group didn't want to belong anywhere, had no opinion, and wanted to wait and see how things would pan out.

The news from the front wasn't very optimistic, and we didn't know what to expect the next day. The big nail in the coffin was the

CHAPTER 7: ZDENĚK ŠKARVADA

Zdeněk Škarvada
- in mines- right

capitulation of France. Obviously, we all became very nervous, the tension was tangible, and it led to open conflict with the head of the camp; Russians reacted accordingly; they forbid school for COs and NCOs, took off all our personal log books, radios, maps and all our personal possessions which meant they took over the control of everything in the camp. Under these conditions, it wasn't a surprise that the relationship between our COs and Russian COs froze.

The Russians turned their attention to us, NCOs, and tried to bring us to their side. They allowed us outside walks, swimming, sports, and musical productions. They even showed us films since they believed we were not as strong-willed as our officers. When spring came and the weather got better, it was pretty depressing to stay behind the wire, so we all tried to get some work outside. I hail from the village, so when I heard they have about 30 horses outside a camp, I welcomed that as a chance to break the monotonous time. I helped to repair the road, transported the goods from the train station to the village, and generally whatever was needed. It was much better than sitting inside and doing nothing."

As one saying says, "All is well that ends well," the Russian internment ended. Zdeněk and others could finally leave Russia on August 12, 1940. They went to Odesa since the Russian ship *Svanetia* was waiting in this port. It took 40 Czechs through the Black Sea to Varna in Bulgaria and Istanbul in Turkey. They left the boat and boarded another called *Sardinian Prince*, which waited ten days before it was incorporated into a big convoy. The journey wasn't very safe since Italian planes attacked the ships daily, and Czech soldiers had to keep anti-submarine and anti-aircraft watches. From Port Said, already under the British flag, they were shipped to Bombay and around Africa to England. Zdeněk got to Liverpool on October 27, 1940, so it took him over a year to fulfill his dream and get to England. Such was his determination to fight.

There were five transports from Russia to England in all:

-The first one left on March 17, 1940, from Jarmolinky across Odessa to France, and 21 people with valid French documents were on board.

-The second left on April 8, 1940, from Oranek across Odessa to France with 46 men on board. It arrived in Marseilles on May 12, 1940.

-The third left on July 7, 1940, from Suzdal across Odessa, Bombay, and around Africa to Britain. Together 75 people were on board the *Narkunda*, and it arrived in Liverpool on October 27, 1940.

-The fourth left on February 16, 1941, and on board the *Cameronia* was 58 people. It went from Suzdal to Odessa and around Africa to Glasgow, where it arrived on July 12, 1941.

-The fifth transport left Murmansk on April 28, 1942, as part of Convoy QP11. On board the cruiser *Edinburgh*, there were 8 Czech airmen. On May 2, 1942, the cruiser was sunk, and leader Colonel Josef Berounský was killed; the remaining seven airmen were saved and returned to Murmansk. Eleven days later, on May 13, 1942, they boarded the cruiser *Trinidad*, but it was sunk on May 15. Czech airman Josef Ferák got killed. The remaining six were saved on May 19. They got safely into Port Greenock.

He was accepted into the RAF and sent to various courses. Then he became a member of the already famous 310th Squadron, where he served as the escort to the convoys. Finally, on February 4, 1942, he returned to the base after a month-long particular course:

"I went by train from Manchester to Perranporth in Cornwall through the night. It was about 11:00 when I hopped off the train and rushed to the base. I informed them about my return and was pleased we had no casualties that month. I was tired after the long journey and was looking for some rest when my friend Leopold Šrom came to me with a smile."

"Zdeněk, please, could you take over my duty? It's only 30 minutes left. I have a week's leave and going to Scotland by train takes

me 24 hours, while by plane, I would be there within two hours. You see, there is a Transport Command plane that would take me, but I have to leave now."

I wasn't very pleased, and before I could say NO, Leopold was rushing towards the plane with his suitcase. The plane's wheels barely left the runway when the siren announced the alert. I swapped the look with the squadron leader, but he told me:

"Sorry, you have a readiness, so there is nothing we can do about that."

Everything was back in the groove, the learned routine worked as clockwork, and I was in the air within a minute with three other planes. We had to patrol above the Scilly Islands, where some motion was spotted. Once in the air, I realized something was wrong with my dashboard. I was disturbed since altitude gauges were disturbed by the unsettled engine. I checked everything I could, but as it seemed, the whole problem couldn't have been sorted out in the air. I called the section leader to tell him about my problems, but I didn't hear anything, and obviously, he couldn't hear me either.

Why? What's wrong, for fuck sake? I checked the gauges from left to right and from right to left. Nobody heard me; I didn't hear anybody or anything since the engine stopped, too. I lost contact with my group, which was far ahead of me now; strange that they didn't care why I was limping behind them. I called again but to no avail.

Okay, that's it. I am done, went through my head, and I opened the hood of the cockpit and unplugged my helmet. Then I unclipped the seat belts and made a back flip. Bang!!!! Spitfire pared the way with me; I went one way, and he went the opposite. I descended very quickly and didn't have much spare time. The parachute got opened, and I saw the water under me. Before I managed to release the parachute, I hit sea level.

I was buried under the canvas, and it took me a lot of energy to open the harness and get rid of the now heavy parachute. While Mae West was blown immediately, the dinghy stayed flat, and the compressed oxygen bottle only made three weak farts, and that was it. I tried to pump it up by hand, but it was tough during the

practice, and now it was even tougher. My hands got tired, and I was cold. My watch stopped at 12:42 GMT, and the world around me turned grey. The last words I remember before losing consciousness were, 'KEEP FLOATING.'

310th Sqn organized the search, and five planes returned to the estimated place. Finding a small yellow point in a stormy sea was difficult and didn't come to fruition. Furthermore, the searchers encountered a lone Junkers Ju-88, and a short but intense battle began. Sadly, the Spitfires returned to base with feelings that their friend was lost for good. However, the guardian angel was working overtime for Zdeněk, and he was half-dead and was found by a German ship watch and picked up.

He was sent to Brest, and from there, he was transferred to Dulag Luft in Frankfurt via Paris. After spending two weeks in solitary confinement, where he was taken out only for interrogation, he was sent to Stalag VIIIB in Lamsdorf. Then he went to Stalag Luft III, where he met many of his Czech compatriots shot down before him. He teamed up with Czech bomber pilot Vilém Bufka. He found a way to improve their menu since Red Cross parcels used to come irregularly.

"Camp food was fundamental, and we would have been hungry without the supplement of Red Cross parcels. So, it was convenient to team up with one or two other Kriegies and share the contents. It lasted longer and brought more variation. I teamed up with Vilém Bufka, a good cook, and I liked cooking too. Once, we got a few boxes of blue cheese to our room, which was well past its expiration time and was why we found it in the dump. It produced an incredible stench, and no wonder almost all the occupants left the room. That didn't bother us. Then, in horror, I opened the lid, jumped away, and put the top back. It was full of worms."

"Sorry, Vilém, how do you want to eat this? I don't have the stomach for that," I told my compatriot.

"He said with self-confidence, "It will need a special amendment," and brought the wash basin we stole from the Germans a few days ago. He prepared a few makeshift pans and brought out the rest of the hard-saved margarine. Then, he lit a fire, opened the boxes, and poured the contents into the wash basin. The

CHAPTER 7: ZDENĚK ŠKARVADA

Zdeněk Škarvada
- With wife after the war

cheese started to melt, accompanied by a strong stench, and worms began to leave the stinky bath. They climbed onto the rim, appearing panicked and unsure of where to go, but Vilém knew what to do and helped them into the fire with a piece of wood. They sizzled out.

When no others appeared, he poured the yellow liquid into the pre-greased and prepared pans, then placed them behind the window to cool. After about an hour, the mass had evaporated, leaving what looked like proper gruyère cheese. We weren't worried that someone else would steal and eat it. We brought bread and salt and ate most of one pan, which was very well-edible."

From 1942 to 1944, he was transferred between many camps, including Stalag Luft IV Memmel and Stalag Luft VI Gross Tychow. The conditions on the coal ship, *Insterburg*, were inhumane, and the 800 POWs on board were convinced that the ship would be sunk at sea. What followed went down in history as the "Run-up-the-Road"

The ship's crew appeared to be Russians controlled by the German Navy. The Insterburg was escorted by an E-boat that circled the boat continuously during the voyage. As a result, many prisoners of war suffered from seasickness. The air was thick with diarrhea, vomit, and body odor, with no ventilation or fresh air. The prisoners were kept in complete darkness for three days, with no food or water from the Germans. Only when they could climb the ladder and get on the deck for a short time could they breathe fresh air. The heat inside the ship was oppressive, and the prisoners could only wet their lips with water.

After three grueling days, the ship finally reached shore, and the prisoners were ordered to climb out to the deck. It was challenging to climb the ladder after such an experience, as their bodies were weakened. In addition to their regular guards, the prisoners

noticed many German Navy guards that were heavily armed and accompanied by dogs. The guards looked young, which the prisoner assumed meant they were from the Hitler Youth Movement. This added to the intimidation, and things got out of hand when the arrival coincided with a daylight air raid. The chaos lasted 40 minutes until the siren announced the raid's end, and things returned to normal. The prisoners were made to wait in sweltering heat for another hour before being ordered to board the train for a 75-mile journey to Gross Tychow. Zdeněk recollects:

"German commander Pickhardt addressed us as Luftgangsters and Terrorfliegers. I believed his entire rant was to only agitate and infuriate the young German Navy guards. After speaking, he gave a signal, and we were forced to jog from the railway station through the village. The local residents spat, jeered, and waved their fists at us, trying to encourage the guards to make our lives even more unbearable.

Once we were out of sight of the village and on the road among the trees, the guards instantly attacked us with bayonets, rifle butts, and dogs. We were forced to stampede, with the dogs snapping at our heels, and we tried to avoid the jabs from the bayonets and swings from the rifle butts. The guards pulled out knives and cut away our packs, knowing they could take anything they wanted. They showed no mercy and seemed to be enjoying themselves. Our bodies, already hungry, thirsty, weak, and exhausted, could have done without this brutality. Anyone, who had fallen by the wayside for exhaustion, has been hit by the rifle's butt or jabbed by bayonet. After about two miles, the stampede began to lose its momentum. Many Kriegies had either collapsed or were lying seriously injured on the road. We realized that Pickhardt wanted us to run away and had set up machine guns hidden among the trees every few hundred meters.

When we finally arrived at the gate of the POW camp, I collapsed on the grass, trying to catch my breath. It took about an hour for the last limping Kriegies to arrive. We were then told that we had no permanent accommodation yet, as the entire camp was still under construction."

CHAPTER 7: ZDENĚK ŠKARVADA

This ship took him to Gross Tychow

Zdeněk Škarvada remained there until the guns from the approaching Russian front could be heard daily. Then, on January 6, 1945, during roll call, the Kriegies were informed about taking their most essential possessions and leaving the camp for the railway station to be taken to the interior of Germany.

Fifty-five years later, Zdeněk remembered the details vividly:

"When we left the camp gate, no one knew what lay ahead. We carried as much as we could and were each given half a Red Cross parcel for the journey. The weather was terrible and very frosty. After daily walks of up to 30 kilometers, we slept in barns or sometimes even on branches of the trees in the snow. The attacks by Typhoons or Mustangs made our miserable life even worse. They often mistook us for Germans and strafed our column. We were herded into a large barn during such an attack. I was there with Franta Petr (311th Sqn Czech airman), and unlike the others on the ground, we climbed to the first floor. Suddenly, we heard the sound of a Mosquito engine and then the whiz of bombs being dropped. One exploded very close to us, causing the entire barn to jump into its foundations. It was about 01:00 and nobody slept; all were up and moving.

The Mosquito made another attack, and the trace marker bullets flew overhead. After the first attack, eleven dead bodies were on the hay, and many were wounded. The bullets also set fire to the

Author with Zdeněk
Škarvada 2008

shingle roof, causing everyone to want to escape the burning inferno. When we got out and stood in the cold mud, we realized our boots and clothes were left inside. Our journey would have been impossible without them, so we had to return to the burning barn. "Lucky we managed to salvage our things, and within ninety minutes, the whole barn turned into piles of charcoal."

The column arrived in Hannover, and the American army freed the Kriegies on May 2, 1945. After many years behind the wire and a 1,000 km long trek, they were finally free. The feeling of freedom almost paralyzed them, but they were happy to have survived and no longer had to walk. The feelings of pure joy came much later when everything sank in.

Within two days, Zdeněk was flown to England with others. He underwent medical examinations and was dispatched for a month's leave. He returned home in August of 1945. He was sent to teach at the flying school in Hradec Králové and Pardubice. In 1950, he was summoned to the Ministry of Defence, where he was forced to sign his resignation and was released from the army.

Not only was he jobless with no income, but he was also demoted to private. Forcibly, he was also moved from his flat in Brno to one in Jevíčko in Moravia, where he lived with his wife and two sons. Since "Westerners" were only offered jobs in mines or cooperative farms, Zdeněk chose to work in the mines. He started as an ordinary miner near Česká Třebová. Still, he soon developed a liking for working with drilling machines and became an expert. He was dispatched to school and in 1964 was forced out again from his living quarters and had to look for another flat and job. He, therefore, was forced to Ostrava, where he was immediately employed in a special exploration team in a deep mine. He was also invited to join a newly formed yachting club for the miners, which enabled him to sail in the Baltic or Adriatic Sea.

A year later, in 1965, during a political thaw in Czechoslovakia, Zdeněk was partially rehabilitated, and his rank of major was restored. After 20 years, he fulfilled his dream of boarding a plane again. Although he retired, he worked in the mines as a language expert and dealt with foreign contractors. He then moved on to metallurgy. He physically retired at the age of 73. After the Velvet Revolution, he was fully rehabilitated and survived the Millenium in excellent spirit. Along with Otakar Černý, he became the only remaining Czech POW.

8 Chapter

BOHUMÍR FÜRST
Family Friend

Bohumír Fürst

Bohumír Fürst was born on October 1, 1909 in Opatovice, Moravia. His parents were poor and worked in a quarry, but Bohumír wanted more for himself. Despite the lack of money for education, he was accepted to the School for youth pilots in Prostějov, where he trained as a pilot and later became an instructor.

After his country was occupied, Bohumír left on June 22, 1939, and escaped to Poland. He had two options: either stay in Poland and join a group led by General Ludvík Svoboda or sign a five-year contract with the Foreign Legion. He was hesitant, as he knew what to expect, but he had a minimal choice. So, on July 26, 1939, he boarded the ship *Castelholm* and set sail for France along with 190 other airmen.

Upon arriving in Marseilles, the airmen entered the home of the Foreign Legion, the fortress of St. Jean. From the beginning, it was clear that they were in for a tough time. Bohumír recalled that they were forced to hand over all their documents and change into

CHAPTER 7: ZDENĚK ŠKARVADA

This ship took him to Gross Tychow

Zdeněk Škarvada remained there until the guns from the approaching Russian front could be heard daily. Then, on January 6, 1945, during roll call, the Kriegies were informed about taking their most essential possessions and leaving the camp for the railway station to be taken to the interior of Germany.

Fifty-five years later, Zdeněk remembered the details vividly:

"When we left the camp gate, no one knew what lay ahead. We carried as much as we could and were each given half a Red Cross parcel for the journey. The weather was terrible and very frosty. After daily walks of up to 30 kilometers, we slept in barns or sometimes even on branches of the trees in the snow. The attacks by Typhoons or Mustangs made our miserable life even worse. They often mistook us for Germans and strafed our column. We were herded into a large barn during such an attack. I was there with Franta Petr (311th Sqn Czech airman), and unlike the others on the ground, we climbed to the first floor. Suddenly, we heard the sound of a Mosquito engine and then the whiz of bombs being dropped. One exploded very close to us, causing the entire barn to jump into its foundations. It was about 01:00 and nobody slept; all were up and moving.

The Mosquito made another attack, and the trace marker bullets flew overhead. After the first attack, eleven dead bodies were on the hay, and many were wounded. The bullets also set fire to the

Author with Zdeněk
Škarvada 2008

shingle roof, causing everyone to want to escape the burning inferno. When we got out and stood in the cold mud, we realized our boots and clothes were left inside. Our journey would have been impossible without them, so we had to return to the burning barn. "Lucky we managed to salvage our things, and within ninety minutes, the whole barn turned into piles of charcoal."

The column arrived in Hannover, and the American army freed the Kriegies on May 2, 1945. After many years behind the wire and a 1,000 km long trek, they were finally free. The feeling of freedom almost paralyzed them, but they were happy to have survived and no longer had to walk. The feelings of pure joy came much later when everything sank in.

Within two days, Zdeněk was flown to England with others. He underwent medical examinations and was dispatched for a month's leave. He returned home in August of 1945. He was sent to teach at the flying school in Hradec Králové and Pardubice. In 1950, he was summoned to the Ministry of Defence, where he was forced to sign his resignation and was released from the army.

Not only was he jobless with no income, but he was also demoted to private. Forcibly, he was also moved from his flat in Brno to one in Jevíčko in Moravia, where he lived with his wife and two sons. Since "Westerners" were only offered jobs in mines or cooperative farms, Zdeněk chose to work in the mines. He started as an ordinary miner near Česká Třebová. Still, he soon developed a liking for working with drilling machines and became an expert. He was dispatched to school and in 1964 was forced out again from his living quarters and had to look for another flat and job. He, therefore, was forced to Ostrava, where he was immediately employed in a special exploration team in a deep mine. He was also invited to join a newly formed yachting club for the miners, which enabled him to sail in the Baltic or Adriatic Sea.

winter uniforms. Their civilian clothes were sold to a local pawnbrokers. They were forced to carry sand by hand back and forth from one corner of the fortress to another.

After three days, the airmen were herded onto a ship bound for North Africa. They arrived in Algeria on August 23, 1939. They were taken straight to local barracks belonging to the Foreign Legion and then transferred to Sidi-Bel-Abbés. Despite their hope for food, they were only given medical examinations and were sent to bed on an empty stomach.

The boot camp was hard, mainly due to the extreme heat. The officers were Germans who showed no mercy. When the war was declared, the airmen demanded to be released to join the Air Force, as they had escaped their country to fight, not to carry stones from one place to another.

Bohumír continued:

"When we complained, our sergeants let us clean rooms infested with lice or sweep pavements in the town. Midday temperatures reached 54 C, and we still had to wear our winter uniforms. Once, we had to undergo a 50 km tour. We set off early morning and went to the mountains of Tesla. After getting there, we had lunch and set off on the return journey. With a lack of water and in winter uniforms, which we couldn't take off, the results were tragic. Out of 29 soldiers, 18 were reported sick. It wasn't possible to escape since there was only one way across to Spanish Morocco, which was 200 km away."

Thirst was never-ending, and we could only drink warm water. The availability of the wine was plentiful, but we got half a frank per day which gave us enough money at the end of the week to buy one liter. When France finally felt the lack of qualified pilots, we were sent to Blida, about 50 km from Algiers. Before we could start flying, the cable arrived, and we had to urgently go to Chartres in France. With pleasure, we waved goodbye to Africa and Algiers and were excited that we would get to fight. We were accommodated well and kept our barracks tidy, but French soldiers had dumps inside and around them.

The latrines stood outside, and they were so shattered in the morning that it wasn't possible to use them. The winter was pretty cold, and we burnt anything we could put our hands on. One local builder had forty pieces of the beam under the window of our barracks, ready for spring weather. All have gone without a trace by that time. When we got four months' back pay, we went shopping and bought some presents for home since we believed we would soon be back. If we knew what lay ahead of us…?

I had a case full of gifts, but when we were suddenly sent to the front line, I left it with the laundress, who used to wash our laundry, to pick it up later. Of course, I have never seen the case or the laundress again. After weeks of "phony war," Germany invaded Holland and Belgium, and we were called into action. I was flying in the Chasse Group II/2, and we headed towards Channel. Soon we were ambushed by ME 110, and the group disintegrated into individual dogfights. I spotted two enemy planes and got behind the tail of the second one and gave him a short burst. I didn't have much time to see whether he got shot down. After that, I lost touch with my group and was pretty low in that scuffle.

I pushed the control stick, but my rudder was "lazy" and didn't react, so I returned to base. But it wasn't meant to be an easy way out since I spotted another enemy plane, a Henschel 126, which was taking pictures and monitoring the situation underneath. He didn't know I was around, which gave me a much-needed advantage, and without much fuss, I got close to him and gave him what he deserved. He went down in a spiral engulfed with black smoke. When I landed on the airfield, I felt great; the first real fight and the first shot enemy. I felt a great deal of satisfaction. When I exited the cockpit and went around my plane, my smile froze since I counted 27 holes and some hits in the rudder, which explains its "laziness."

As the war progressed, the chance to get spare parts were smaller, and I had to fly with a different plane every time. Since I got some respect for landing with a damaged plane, a commanding officer gave me worse and worse planes with the conviction that I would land with it anyway.

When I was sick of that, I asked him for the worst possible plane and got it together with my mechanic. Then I announced this would be my plane, and I would not swap it with anyone.

Since spare parts became even more challenging, mechanics often picked up a car, went from airfield to airfield, and tried to salvage any parts.

On June 15, 1940, we had a patrol above Chaumont. When we were returning to base, we spotted German troops barely 50 km from us, so no wonder when we landed, the order to prepare for evacuation was given. We had a few minutes to pack our belongings while the mechanic refilled the plane. When we brought our cases, I realized there was not much space in the cockpit, so I took only the necessary hygienic tools and put my case onto the lorry with others, paid the landlady the rent, and was ready to take off. No need to say that I have never seen my case again.

To show the chaos in which the French army withdrew, I can only say that we changed five airfields in three days. On June 17, we finally landed at Lyon airport, where we were told that France had capitulated. So we Czechs were again in a situation where we didn't know what to do. Since we didn't want the French Higher Command to hand us over to the Germans, we persuaded our Commander to let us fly out to Montpellier and then to Africa again. But once we landed in Montpellier, another order halted us from continuing our desperate escape from German talons. All hopes were dashed, but we didn't give up.

I and eight others obtained permission to command a Lockheed American plane and fly close to Spanish borders. We drove to Port Vender to catch the last ship cruising to Britain. We had the same feelings as one year previously when we escaped our own country and lived in permanent danger of falling to German hands and being punished as "Traitors of the Third Reich." The threat was almost tangible since German troops rolled over Europe and were everywhere, and we had nowhere to go. A spark of hope was struck when we heard England might accept us."

Bohumir Fürst finally got to England, and after necessary medical examinations and formalities, he was accepted into the RAF. Altogether, 35 experienced pilots were sent to flying courses and were taught how to fly Hurricanes. Since none spoke English more than a few words, they were advised to befriend some English girls

Former Bata shoe company in Cambridge 2006

who would teach them how to master Shakespeare's tongue day and night. Bohumir followed this advice and soon was seen with a member of WAAF called Brenda.

On August 17, 1940, the Czech fighter 310th Sqn was announced as fully operational. It was sent to patrol above England, later becoming a substantial part of the RAF in the Battle of Britain.

What was it like put Bohumir in his memoirs:

"We were up in the air three or four times daily and flew at 20,000-30,000 feet. Apart from flying, I tried to learn English as much as possible, and if the weather didn't permit us to fly, we played darts or cards. On September 3, 1940, I shot down my first German. Another success came four days later. We heard the "scramble" and got above London on 20,000 feet. We soon spotted invaders and picked individual planes."

My section leader got behind one Me110 and quickly sent it to the ground in smoke. Then, another aircraft got behind the tail of my number two, so I came to rescue him and gave the bastard what he deserved. What a joy!!! But since bombers immediately followed them, I didn't have time to follow my victim up to his bitter end, so it was not a confirmed one but only a probable one. Brits were very precise in this evidence.

CHAPTER 8: BOHUMÍR FÜRST

Since we were entangled with fighters, bombers managed to drop their bombs load in the London area and turn back to the Channel. We didn't manage to trace them but met the second wave of Heinkels accompanied by Me109. One, the probably young and inexperienced pilot, thought we were Germans going homewards and friendly waved at me with his wings and then showed me his white belly with crosses. Just as if he was on a sightseeing flight, he got in front of me, so I quickly put him into the cross, gave him a merciless burst, and chopped off his wing. He dropped down as a stone, but a young pilot managed to bale out, so he probably survived the war in England. I felt a bit sorry for him, but the laws of war are different "Kill or be killed."

We had to go home to refill and get new magazines of ammunition.

Since the 605th Sqn had significant losses, four pilots from our 310th Sqn were loaned to them. It was Vladimír Zaoral, Vilém Goth, Raimund Půda and me. We were based in Croydon, where we arrived on 18.10, and a day later, we were already up in the air. When the "scramble" was called, we had to dash in full gear for about 300 meters, where our Hurricanes were taxied. Usually, a small cross-country run can be good for fitness. Still, our flying equipment was airtight, so we sweated like pigs when we got to the cockpit, but once we got to 30, 000 feet, it didn't protect us from getting cold. When I got to bed in the evening, I had a fever and sore throat. The next day the "scramble" was called again, and I felt terribly weak. But I didn't tell anybody if they thought I wanted to leg it or was scared. We spotted two Me 109 heading towards the Channel, but as it turned out later, they were only the ploy that should lure us into following them and getting into the trap. I felt really shitty and legged behind when four Me 109 got behind me. I kicked the plane into the sharp left corner and downed in the corkscrew. That maneuver got all blood from my brain, and I fell unconscious due to lack of oxygen. I have no idea how long it took, but when I came to my senses, I found out I had dropped by 27,000 feet and now was only in 800 feet.

Fortunately, the enemy didn't spot me since I would have been

easy prey for him, but now the sky was clear, the only bit of smoke here and there. I had no idea where was I, I only knew that it was midday and we flew southwards, so I turned my plane to have sun in my back. Within five minutes, I spotted the English countryside and balloon barrages underneath me. Finally, I found a small airfield where I landed and was told where I was.

Unbeknownst to me, our flight landed in Croydon, and since somebody saw me dropping down, he reported that, and next to my name, the word "missing" appeared.

After two hours of rest, I took off and wanted to fly to Croydon. Still, immediately after I started the engine, I felt really sick. It was an uphill struggle to keep the plane in the air, so I kept the railway as the guarding line to take me to the airfield. Once I spotted the airstrip, I landed no matter where and lost consciousness again. I woke up the in the hospital and stayed there for a week with high temperatures. When I returned to Croydon, I was informed that we were sent again to Duxford to our 310th Sqn. Sadly not all returned; Vilém Goth didn't return from the same sortie. On my way to Duxford, I popped over to Czech Air Inspectorate. They were stunned when they realized the airman labeled "Missing, probably killed" was standing there alive and kicking. I had a lot to explain. I told them if someone is pronounced "dead," and he, in fact, isn't, that he will survive the whole war.

At the beginning of the war, I wore a talisman. I was superstitious, but I started to believe only in myself as the war went on. Now I had a powerful inner feeling that I would see off the end of war alive.

When I arrived in Duxford, it was in the middle of preparation for Czech Red Letter Day on October 28. The celebration went very well and was attended by Air Marshal Karel Janoušek and Czech Foreign Minister Jan Masaryk. Sadly, the whole day was marred by a tragic accident. During the flight, the planes of Emil Fechner and Jaroslav Malý collided. While the former was instantly killed, the latter was transferred to the hospital, where he died several months later. The bitter-sweet thing at the whole occasion was that Fechner's mother sent him a letter that wandered many

CHAPTER 8: BOHUMÍR FÜRST

Grenville Road, same place 60 years later, 2000 - author

months across Europe, from base to base, followed by its addressee until it finally reached him on October 28, one hour too late. So it was put onto Fechtner's coffin, and it went to the grave with him.

In November, the weather was rainy and foggy, so we were most often grounded, and Germans also didn't fly above London as much as they used to. So I had more free time on my hands, and in Cambridge, I befriended the Formánek family, whom I used to often pop in for a chat or something to eat. They were Czechs, and Mr. Formánek worked for Baťa company in Cambridge, producing gas masks.

"When we were ready, we trained for various types of attacks. During one such flight on December 12, 1940, at 20,000 feet, I caught a bad cold and developed an inflammation of the middle ear. As a result, I lost my hearing and had to stay in bed for two days.

On December 14, President Dr. Edvard Benes visited our base and decorated the pilots with the Czech War Cross. I was one of the lucky recipients, but standing in the cold weather worsened my inflammation, and I was taken to Ely Hospital. I had stayed there until January 7, 1941, when I was discharged. During that time, I couldn't hear anything, and I must admit, I had different plans for a romantic Christmas.

During my recuperation, I went to London. I narrowly escaped a direct bomb hit in a tube station called "Bank" during a heavy bombing. After that, I went to the Czech Club, where I met a member of the exile government from my hometown. He told me how Czech relatives of airmen were jailed and sent to concentration camps. I had no news from or about my family since my emigration two years earlier. It wasn't until after the war that I learned my mother and brother spent three years in a concentration camp. The Germans found the information in the forms we filled out after escaping to Poland, which they managed to obtain during the Blitzkrieg."

Bohumir Fürst married a Czech girl named Marie, whom he met at the Czech Club in London. She and her sister Anna left Czechoslovakia to work in London before the war, and the war prevented their return. In the spring of 1944, he had another close shave with a V1 missile.

"Hitler wanted to turn the tide of the war and sent V1 and V2 missiles over London and other British cities, causing many people to evacuate to the countryside. My wife was away with our daughter, so I was alone in our London flat. One evening, a V1 hit the target about 100 meters from our building, and the pressure wave threw me out of bed. After that, I started going to the U-tube shelter almost every night. I was impressed with the peace the London citizens maintained during the bombing.

Fürst was working as a flying instructor and was ordered to go to Canada to teach. Still, he declined and asked to be relieved from flying duties temporarily. He was mentally exhausted from escaping Czechoslovakia, his long journey to France and the Foreign Legion, his escape from France, the Battle of Britain, and nonstop flying with pupils. He later became the commander of the 510. Courier Squadron and welcomed the end of the war there.

On June 6, 1945, he flew to liberated Czechoslovakia and visited his wife's parents. His father-in-law was shocked to learn that his daughters, whom he hadn't seen for six years, were married to Czech airmen, even more so, that he was a grandfather. After three days in Prague, Fürst returned to London. By August, he was permanently deployed in Prague, followed by his wife in October. This marked the end of his happiness."

In February 1949, he was discharged from the army without any clear explanation. He was only told that he was considered untrustworthy because he had served in a Western army. Despite receiving French and British medals for his bravery in the fight against the Nazis and for helping his country, those who had stayed at home and not fired a shot were now in power and were able to destroy the lives of people like Bohumir.

He tried to find a job and applied at around 20 places but was rejected everywhere. Finally, after a month of searching, he secured a position as a laborer in a metallurgical factory with a gross salary of 2,000 crowns. This was the only income he had to support and feed a family of four. On January 10, 1950, he was arrested in his flat in Prague, and had been taken for interrogation. Later, he was sent to a forced labor camp in Mírov, without any reason given. A month later, the arrest was justified due to his refusal to work.

His appeal was dismissed and he was degraded to the rank of private. While he was in jail, his family - wife and two small children - had no source of income and were kicked out from their flat without any compensation. They were sent to an uninhabitable grain warehouse but could not stay there, so they had to live with relatives in the village. When Bohumir returned from jail, they had to live in an emergency apartment in the attic.

His health deteriorated due to the harsh conditions in prison, and he was taken to the hospital with ulcers and muscle rheumatism. He was freed, after being confirmed he was wrongly jailed. He was given a job as a warehouseman in the Kovofiniš factory in Ledec nad Sázavou. Despite this, he still had to pay for his stay in the forced labor camp. He worked until his retirement, and in April 1965, he was partially rehabilitated, and his rank was returned to him.

Sadly, Bohumir died from a heart attack on January 2, 1978.

Author's note:

You may have noticed the surname Formanek in the previous recollection. That was my grandfather. He used to work as a sales manager for the Baťa Shoe Company, and from 1937-1939, he, his wife, and my 5-year-old father lived in Holland. My father used to tell me his story when I was a small child, which planted the seed of my interest in the RAF.

His story went like this:

"We used to live in the Dutch city of Eindhoven, and at the outbreak of the war, Baťa decided to send about 250 of its employees worldwide. We lived in a family house at 3 Greville Road, Cambridge, and my grandfather worked in the Baťa factory. The stay in England was supposed to be temporary, with the final destination being Brazil. My parents received visas on April 16, 1940, and they bought big trunks, recommended clothing, and waited for embarkation. We were supposed to sail inside the convoy. Still, our transfer to Brazil was canceled since the submarine war started to have significant casualties.

On the streets of Cambridge in July 1940, my parents first met Bohumir Fürst, Karel Šeda, Rudolf Zíma, and Miroslav Laštovka, all from the 310th Sqn. Later, Mr. Laštovka became a navigator in the 311th Sqn. My parents invited them over for dinner, marking our friendship's beginning. During the war, it was common for families to invite airmen into their homes and help them escape the loneliness and fear of being killed.

As a result, the airmen received the Czech cuisine exceptionally well, particularly the Czech specialty of pork, dumplings, and sauerkraut. Bohumir Fürst was a frequent visitor, but on the one occasion when my mother made plum dumplings, all three came, and the entire batch of 100 disappeared in their stomachs. He enjoyed helping my mother in the kitchen, washing the dishes, and sharing stories about dogfights, what it felt like to trace enemy planes, and the sadness of watching a shot-down friend fall from the skies. He also told us about how he was shot down in France, and how he shot down his first enemy plane.

Thanks to Bohumir, our family was invited to Duxford to celebrate October 28, 1940, a significant day for Czechoslovakia. Our president, Dr. Edvard Beneš, decorated him and even gave me a photograph signed by the president. A few days later, he visited and informed us about a tragic accident on the airfield; Emil Fechtner had a mid-air collision with Jaroslav Malý and was killed. He brought me a piece of the propeller from Emil's plane.

War Christmas was always a sensitive and sad time for soldiers, and Bohumir was no exception. He brought his friend Rudolf

CHAPTER 8: BOHUMÍR FÜRST

Grenville Road, 1940 - L-R Miroslav Lastovka with wife, young Leopold Formánek with mother, Bohumír Fürst

Zíma; it was our last time together. That evening, Bohumir left early as he missed his family greatly. On another occasion, he brought his friend, a fighter pilot named Stanislav Plzák, who flew with the British 19th Sqn based in Duxford. I remember Mr. Plzák well, especially his vivid accounts of the war's atrocities. He told me that he always carried a gun on patrols over France in case he was shot down, as he didn't want to be captured by the Gestapo and preferred to shoot himself dead.

One day, Bohumir visited alone and informed us that Stanislav had been shot down over La Manche on August 6, 1941. Both men were among the top ten Czechs in the Battle of Britain. Eventually, we moved to Maryport. Messrs. First, Šeda, and Zima were assigned to various squadrons, and our contact with them was lost. Bohumir spent a week's holiday with us and even invited us to London for his wedding, where I served as a young bachelor. We lived in East Tilbury then, and my father worked at Baťa's factory producing gas masks. The last time Bohumir visited was in 1944, when he brought his wife and young daughter. We returned home as repatriates on September 2nd, 1946, and all contact with our airmen was lost. After returning from the West, my father was sent to work in the mines, and I was denied university admission. By pure chance, I met Mr. Fürst in 1965 during a business trip to Ledeč nad Sázavou, where he lived. In the evening, I was invited to

Author's father- Leopold Formánek as young bachelor on B.F. wedding

his home, and he told me about the horrors he experienced during the communist times in the 1950s."

Authors note:

"My memories of Mr. Fürst are the best and fondest that my parents and I have from the war. He was courageous, kind, and loved his family and country. The story could have ended there, but it had a continuation, and I was glad to play a part in it. I felt that the entire cycle was complete.

I became very interested in the RAF in 1988 and found Karel Šeda from the former 310th Squadron. He came to visit my father, who was 57 years old at the time, after a long 47 years. Unfortunately, my father, who was already ill, died a few months after their meeting. I became friends with Jeff and Marie Carter from Ely, and we stayed in touch through letters and visits until they died in 2015. I loved them, and they appreciated my interest in the RAF and my family's story. During one of my visits to Ely, they took me to 3 Greville Road, where I met the current owners and took a picture at the same spot where my father stood 65 years ago. The owners told me they remembered those who bought the house from my grandfather.

During another visit to the UK, my girlfriend Eva and I were taken to the former Baťa HQ and factory where my grandfather used to work. A nobleman owned the house, but we still dared to knock on the door. When we told him our story, he let us in and gave us a tour of the grounds. During this tour, he told us how much he liked Jan Masaryk and Czechoslovakia, which was very flattering.

I still have all the photos my father gave me of his time in England, various gauges from a Wellington bomber, a pilot's helmet, and various magazines and books from 1940. When I hold these items, the entire film of memories starts to play again. I would never part with them."

Chapter 9

IVO TONDER
The Great Escape and Beyond

Ivo Tonder and his trade mark - pipe

Ivo Tonder was born on April 16, 1913, in Prague. He studied for three years at Lycée Carnot in Dijon before returning home and then at the Technical University. In 1936, he entered Flying school. After the Nazis occupied his country, he left on December 14, 1939. He escaped with friends via the Balkan Route to Slovakia and Hungary. They arrived in Györ after a 50-kilometer walk and hired a taxi to Budapest. On Christmas Eve, 1939, they entered the French Embassy in Budapest and obtained a French visa. Then, they boarded a train to the Yugoslavian border. They quickly crossed and arrived in the village of Subotice and eventually reached Belgrade, where the local Czech colony took care of them.

After a few days of rest, they bought train tickets to Greece and continued to Turkey, Syria, and Beirut. Finally, they were accepted into the Foreign Legion in Beirut and were shipped on the vessel "Patria" deck to Algeria, Marseilles, and Agde. His companions were Zdeněk Hanuš and Karel Sláma, who later became members

of the 311th Sqn and 313th Sqn, respectively. After France surrendered, Tonder boarded a ship and crossed Gibraltar to reach England. He was accepted into the RAF, and he was sent to the 6.OTU in Sutton Bridge for training. He was later dispatched to the Czech 312th Sqn as a pilot.

On June 3, 1942, the Czech 310th Sqn, 312th Sqn, and 313th Sqn were sent to accompany Hudson bombers over the port of Cherbourg. Each squadron flew in formations of three groups. While changing shape over the target, the group was intercepted by 50 Fw190 and caught off guard. As a result, Bedřich Dvořák from the 312the Sqn was shot down. The plane that took him down was in front of Tonder's indicator. Tonder opened fire, but the German was quicker and went upwards. Tonder didn't want to back down and pushed the revs to the maximum to keep the enemy's tail. He opened fire from all his cannons, dramatically slowing down his speed and causing him to fall into a corkscrew. After regaining control, Tonder found himself among four Fw190s. The group leader was before him, and Tonder opened fire without much aiming. Almost immediately, black smoke appeared, and the plane rolled down toward the sea level.

Tonder's number two, probably a young and inexperienced pilot, accompanied him, and Tonder got entangled with the remaining two enemies. The dogfight had no winners or losers, and Tonder suddenly realized he had run out of ammunition. Luckily, the same happened to the two FW 109s, and the fight stopped immediately. The Germans honored some WWI chivalry and waved to Tonder to appreciate his skills. Tonder was relieved that his plane had no visible damage and set the course for England. Suddenly, he heard a big "bum, bum" sound, like someone was shooting cannons at him.

He recollects at this moment:

"I looked in the mirror and quickly checked to my left and right but saw no one. Looking back, I saw that my cockpit was filled with smoke. Everything was happening in mere seconds. I subconsciously thought I was on fire, the biggest fear for airmen. My brain gave me the order to 'Get out,' so I quickly unbuckled my seatbelt,

opened the hood, pulled out the radio cords, and jumped away. During my descent, I thought it would be better not to open my parachute too early. I saw my Spitfire flying forward at full speed, with no smoke or fire, from the corner of my eye. I was perplexed. I had heard the bang, seen the fire flash, and visited the smoke, but there were no enemies in sight, and now my plane looked unscathed and happily flew forward...without me.

The only logical explanation was that two small charges under the radio, which were there for the case of belly landing on enemy territory, had exploded, resulting in the flash and smoke. I acted automatically, knowing I had to get rid of my entire parachute before reaching the water level. I inflated my Mae West and got the dinghy out of the parachute cover, which I easily inflated by opening the valve.

I was convinced a British rescue plane would pick me up soon, so I set my course for England and started paddling as fast as possible. After four hours, I heard a droning of aircraft, but I didn't know who it was. I prayed it was ours, but it was only wishful thinking, as two FW 190s appeared above me. They spotted me and started circling, and one flew toward the French coast while the other stayed above me.

Soon, a big Heinkel He 59 with two large floats was heading toward me. I had colts in my boots, so I pulled them out and hid them behind my paddles. I intended to shoot the crew dead and fly away, but my plans were dashed, when an armed soldier, with a machine gun, stepped on the float and aimed at me. Reluctantly, I dropped both colts into the sea. I didn't know where I was flying to, but soon I found myself in a hospital in Cherbourg. They gave me dry trousers and a shirt and removed my drenched uniform. I was invited to dinner and driven to the Luftwaffe mess. I was starving, so I didn't decline the invitation. The airmen were friendly, so I felt comfortable, and after being driven back to the hospital, I slept like a log. The following day, I was given back my dried and ironed battledress and taken to Dulag Luft in Frankfurt am Main."

Ivo stayed there for four days, and then he was sent, along with 20 other prisoners, to Sagan by train. He, of course, wanted to

Readiness Ivo Tonder on the left

escape during the journey, but his attempt was short-lived. He then arrived at Stalag Luft III in Sagan. Not long after, he was introduced to the Escape committee and got actively involved in digging the tunnel. When Roger Bushell arrived in Sagan from Prague, where the Gestapo held him, he hated the Germans even more. He became the Big X in planning the escape, The Great Escape. While digging, Ivo didn't forget to try and escape, and he took every opportunity to do so.

On April 1943, he and Australian F/Lt Cornish blended in among Russian prisoners who were sent to chop down trees in the northern part of the camp. They dressed like Russians, so blending in wasn't a problem. However, Ivo overheard a guard shouting at another, "Two Brits among Russians." As he understood German, he decided to leave the group and hide before he was discovered. He hid behind the barracks and took off his Russian uniform. Still, unfortunately, his partner was too far away to be warned and was eventually discovered and spent two weeks in solitary confinement.

Desmond Plunkett later informed Ivo that a pile of empty cans was in the middle of the camp and that a cart with horses came to take them away. Ivo obtained two sacks and planned to hide inside

them, cover himself with the cans, wait until morning, and escape camp. The plan succeeded, and Ivo and Plunkett stayed under the load until morning. However, the waiting was more agonizing than expected. Finally, in the morning, the Germans decided to fill the trench dug around the camp. The whole cart was taken and dropped its load, not expecting to find two prisoners underneath.

When Ivo returned to the camp, he was told that the digging work on the tunnel "Harry" had resumed. He was teamed up with John Stower, and they were digging so efficiently that they were given slots in the "Great Escape" for their merit. The digging shift lasted four hours and was physically demanding.

Ivo obtained false documents as a Czech worker from the Protectorate who worked for Focke-Wulf. At the same time, John Stower pretended to be a Spanish forced laborer who spoke Spanish fluently. When Harry was finished, Ivo climbed down for inspection and admitted feeling claustrophobic.

D-Day was set for the night of 24th/25th March 1944, and Ivo remembered the dramatic and nerve-wracking moments. The first people to enter the tunnel were John Marshall and Arnošt Valenta, who were to dig out the last few inches of soil and open the entry lid. Behind them came Bull, Mondschein, Dowse, Krol, Stower, and Ivo. Behind them were the people who operated the air pump and the experienced diggers who organized the shuttle carts. Unfortunately, so many people consumed more oxygen than the pumps could produce. Moreover, everyone was nervous because the exit lid couldn't be opened. When it was found out that the tunnel was too short, Bushell came up with a rope signal, which was a good solution in the given situation. Still, it, along with other misfortunes, slowed down the escape of the prisoners.

Ivo and Stower operated at "Piccadilly" station and then at "Leicester Square" station. They had to assist twenty people before they could leave the tunnel. Major Dodge entered the tunnel for the first time and suffered from claustrophobia. Before he could bypass Ivo and pull up the shuttle cart, he almost kicked Ivo to death. Ivo's other customer was a man named Bretell. When he arrived, the air raid sirens sounded, and the Germans turned off the lights.

The darkness inside the tunnel was palpable. Bretell didn't panic. He carefully bypassed Ivo and pulled up another cart. Still, it either got jammed or Stower on the other end didn't want to let it go, resulting in the rope breaking. Instead of tying the rope, Stower let it drop coiled, and Ivo had difficulty untangling it and tying the loose ends.

Ivo recollects:

"I felt it took hours to do it, and when I met my partner John Stower, he annoyingly informed me that we missed our train and would have to go on foot. When we finally got out, we had to change our plan. Our food supplies were minimal, only some chocolate since we calculated with a train ride. Hence, we were supplied with German food coupons.

We hid in the woods when dawn appeared and slept the whole day. The next night we headed towards the Czech border. Still, we were somewhat surprised by the number of people sent into a man-hunt after us, including escaped soldiers, Protectorate policemen, young boys- everybody. Walking during the daylight was impossible, so we made a makeshift shelter from the snow and were so well hidden that even bypassing deer didn't mind us. Then I spotted a gamekeeper with a rifle, so I woke up Johnny, and we hastily discussed what to do. We decided to go deeper into the wood, and on the way there, we passed a homestead on the edge of the small village.

Unfortunately, some kids spotted us, and the dog started barking. Kids ran to the first house, and three men immediately ran after us. We ran away and headed toward the wood. It was a real run for life. Exhausted, we got there, and after short rest, we shaved so as not to look like highwaymen. Finally, we concluded that to try and cross the borders was nonsense. We had to get away from them. So, we got to the train station, bought the ticket, and headed towards the Baltic Sea to Szczecin port. But sadly, we had no luck finding helpful sailors who would let us get on board.

Very annoyed, we bought tickets for a train going southward toward Zittau. Totally starving, we went to the railway station restaurant and bought ourselves a beer and soup. Then, we bought tickets

CHAPTER 9: IVO TONDER

Escape from France

to Liberec (a Czech city not far from the German border) and got on the train. The German Kriminal Polizei entered the train and checked the documents. We handed them forged papers and passed fine. But, sadly, one member turned back, looked at Johnny's trousers, and told his mate that they have the same color as those he spotted at the guy who was in their prison. They were made from plum-colored Australian uniforms. So, we had it, that was it.

We were taken to Liberec, where I met another four guys from The Great Escape. I was kept there while the others, including Johnny, were taken away- supposedly back to the camp. I was defending that we came together and I am a British airman, so I must go with them, but to no avail. After three weeks, when nothing happened, four SS men in black uniforms came to my cell and picked me up. I was thrown into the car and thought, I finally go back to Sagan, but I was wrong. They took me to Prague to Pankrác prison. Funnily, they didn't know where it was, so I had to show them the way. Once there, I was put into a cell. There was only one very filthy, uncovered toilet, one collapsible bed, and a table with a chair. I was very thirsty and didn't know where to obtain some water, so I knocked on the door. They replied that I could drink water from the toilet after flushing it, which was worse than expected. After many days of inactivity, I was finally called for the first interrogation in Petchka Palace, near the central station. The guard brought me soup, and the investigating official told me something.

I replied in German that I was an airman officer under the protection of the British Government, so he should be careful in what he intended to do. He backed up and told the guard to bring me another soup plate. After that, he asked me various questions about my squadron and other military-related subjects. I didn't know how much they knew about me or us Czechs in the RAF. Here, I was informed about my 50 killed Sagan friends and that Johnny was among them. In fact, all three who were taken away from Liberec prison and with who I wanted to go were shot. Only I was saved. I don't know why.

I didn't know then that they knew exactly who I was and arrested my whole family. I was kept for seven months in solitary confinement. When the guard or NCOs came to me and demanded that I stand in attention, I replied that I was a commissioned officer. They are non-commissioned officers, so they should stand in attention. They often chuckled, and slowly I built a reputation as a no-bullshit man. One day I stood in the yard, and there came a guard, who didn't know me. He gave me a terrible slap, and I fell over. I started to see red, jumped at him, and ended up on the ground. He tried to pull the pistol from his holster but didn't manage that since I knelt on him. I shouted at him in German that I was an English officer, and he slapped me how he dared to do so. When other guards came in the running and pulled us apart, this guy ran away.

Soon after, I was called to the office. The interrogation officer apologized for his subordinate's behavior, saying it won't happen again. But I still ended up in correction. I didn't know how long I would be there, but I didn't get any food and thought I would die of hunger. I didn't know how many days came until a Czech guard gave me an apple and told me to withstand it since it would soon be over. The day I was released, I bumped into Bedřich Dvořák, who was on the run with Desmond Plunkett. They were arrested

after the mass murder of the fifty, so they were sparred. We were sent to Loreta, a jail near Prague Castle, while Plunkett was soon sent back to Sagan. Bedřich and I were then sent to Barth and Colditz, where I didn't participate in clandestine activities. We were liberated on April 16, 1945."

When Ivo returned to his liberated country, he bought a small farm in Eastern Bohemia, which he later had to sell, and bought a bigger one near Mariánské Lázně in the West of Czechoslovakia. He started breeding hens, 2,000 in all, then also horses, cows, and pigs all by himself. It slowly built up. He got two men to help him. Although it was damn hard work, he was pretty happy there. After the election in 1948, he got a letter from the local town hall, where he was warned that if he didn't join the Communist party, he would be kicked out of the farm. It dawned on him that the future was not for him here, so he decided to escape.

The problem was they had two small children, one still in diapers, so they didn't easily move through the forest in the darkness to the border. The first attempt wasn't successful, and lumberjacks spotted them. They called the police, and soon Ivo and with family were arrested and deported. Since the kid was still breastfed, they showed some mercy and set them free along with the mother, but Ivo was sent to jail and had stayed there for five months without any trial.

When he was released, he frantically started to think about another escape.

He met fellow airman Karel Sláma and told him they could fly away in a nicked plane. All was discussed, the time and place set, and the luggage packed. Still, the weather changed, and the landing strip on the field got muddy, and it wasn't possible to land there or take off, so another attempt was aborted. Soon after Christmas, Karel Sláma was arrested, and Ivo followed him. He was taken to Brno, and history repeated. He entered the interrogation room and received a mighty slap. It irritated him, and he jumped at the policeman who interrogated him, and the fight started. Three more policemen ran into the office, but Ivo didn't give up easily and threw everything that came in hand at them. Finally, he gave

Ivo Tonder next to his plane

up and was taken to the cell among crooks and thugs. He discovered how to escape through the ceiling and started working on it. When everything was almost ready, and he intended to escape, the cell door opened, and the guards entered. He was betrayed by one of the thugs.

Spring passed, summer passed, and autumn and a trial came. He was accused of espionage, high treason, and connection with Western imperialists. Ivo decided to play the game with a judge, so when asked if he was in jail, he replied that he certainly was. The judge took it as a small victory before the jury, keeping it similar.

The next question was about the length of the sentence to which Ivo replied that he wanted the death sentence.

"How come you are still here then?"

A quiet laugh was heard among the audience.

"What have you been convicted of?" asked the judge again.

"For high treason and espionage by Reich court," replied Tonder.

This verbal tug-of-war lasted for a few hours. It resulted in Ivo being sentenced to one year while other defendants were convicted of twenty-five years. Finally, Ivo got to a small jail in Blansko in Moravia. He met a guy who seemed pretty handy and asked if he would consider escaping with him. Once, the guard made a

CHAPTER 9: IVO TONDER

makeshift cinema in the corridor between lines of cells and showed inmates some propaganda films.

Ivo and his mate took this chance and walked past guards confidently as if they were going to the kitchen. Nobody paid attention, so they continued and climbed over the wire. Since they wore prison uniforms, they decided to break into a factory where they usually used to go for shifts and pick up working overalls. They looked like ordinary workers going to or from the shift with bags across their shoulders. They went on foot and got to Brno to Ivo's friends. They heard on the radio that two inmates were on the run and heading toward Prague, so they knew they were fine. They were fed, got a couple of bottles of wine, and continued their journey.

After four days, they reached the River Thaya, still in Czechoslovakian territory. There were fields and a few bushes around them, so they rested in one of them and drank the wine. When they wanted to get up and continue, they spotted gamekeepers with rifles approaching them in a row. They found themselves in the middle of a hunt. As the hunters came closer, Ivo approached them with a bottle.

"Come and join us," he shouted at them and showed them the bottle.

Surprised hunters didn't wait long and took a sip, and their local National Committee leader-chairman asked them what they do here.

Ivo swiftly replied that they were telephone line repairers and they got a call that they had a job here.

Chairman nodded with respect:

"Wow, I reported that two days ago, and you are already here. Good job. Look, we have a party to celebrate the hunt, fuck work today. Come with us, and we all get drunk." So, Ivo and his mate aligned with the hunters and went forward. While gamekeepers were looking for something to shoot, both escapees slowly moved to the edge of the line. Both men left the row and disappeared when the hare and partridges were annihilated. At night, they crossed the border to Austria.

So, Ivo was safe, but his wife was in jail. Ivo's running inmate intended to form some anti-communist resistance net and decided

Meeting after 45 years Ivo Tonder, Karel Mrazek, Otakar Hruby - England 1990

to return to Czechoslovakia. He promised to help and get Tonder's wife and kids out from behind bars. Meanwhile, Ivo was transported to England, where he rejoined the army. But although he was a CO with medals from the war, he was accepted as a new recruit undergoing a boot camp and peeling potatoes in the kitchen.

Along with Ivo, there were more Czech airmen; among them was Tomáš Vybíral, the most decorated Czech pilot. This attitude pissed off Ivo very much. He went to the commanding officer and asked him to be immediately released from R.A.F., from hero to zero in six years.

His wife got a permit to leave the internment camp and took this chance to meet the people who organized escapes and ran to Vienna. But both of their children were still in communist Czechoslovakia, staying with their aunt and uncle. Smugglers picked them up and got them across the River Thaya to Austria on inflated tires, but they were caught since it was in the Russian zone. But as they were Austrians, they were released, but the children were put into some Austrian family. Tonder's friend helped them again.

He outwitted the surrogated parents and took the children away, claiming they had TBC. They were in the hospital from where Ivo's wife's brother picked them up. So, the family was finally together. The whole escaping cavalcade lasted 18 months.

Ivo was 38. He knew nothing, had nothing but his family, and had to start a new life in a different country from scratch. His rich war experience meant nothing. He paid a high price to fight for freedom.

10 Chapter

BEDŘICH DVOŘÁK
Forgotten hero

Bedřich Dvořák

Bedřich Dvořák was born on February 18, 1912 in Chotěnov.

After leaving secondary school, he decided to become a professional soldier. He entered the Military academy and retired as a pilot. He left Czechoslovakia for Poland on June 11, 1939, with four other airmen from his unit - František Fajtl, Otakar Korec, Rudolf Fiala, and Bohumil Kimlička. From Poland, he left on board the ship *Castelholm* to France, where he was trained in French planes. He took part in the Battle of France as a member of Groupe de Chasse III/7, which was equipped with fighters Morane Saulnier MS-406. Before he got injured, he flew 58 operational hours with them. For his war effort, he received Croix de Guerre. After France's collapse, he boarded General Chanza, which took him around Gibraltar to Africa; he got to England on the Neuralgia.

As an experienced pilot, he was a welcome addition to RAF. He was accepted on September 19, 1940, and sent to the Czech 312th Sqn. Unfortunately, on June 3, 1940, he was shot down by Fw190

 CHAPTER 10: BEDŘICH DVOŘÁK

above Cherbourg, and his Spitfire exploded. Dvořák got out with great difficulties, and with the broken left hand, he landed in the sea some 8 kilometers from Cherbourg. He almost drowned because he couldn't get rid of the parachute.

He recalled that after the war:

"I was flying as a leader of the second section, and via radio, we got information about enemy fighters ahead of us. So the whole formation changed direction. In swapping positions, I suddenly felt the pain in my right leg and heard an explosion behind me. I wanted to turn right, but my rudder didn't react, nor did the control stick. My plane got into straight dive, and I couldn't control it. I tried to inform the base by radio that I would bale out, but the radio didn't work either. So I opened the cockpit, undid the seat belts, unplugged the radio, and removed the helmet. But air pressure kept me inside the cockpit, so I pulled the ripcord and opened the parachute.

My left hand got trapped in the ropes and was broken in two places. Due to that, my parachute didn't open as it should, and it looked like a sack, resulting in my falling down at great speed. At about 6,000 feet, it finally opened, and I spotted three big holes in it. Shortly after, I hit sea level while my plane hit the ground and exploded near the shore lighthouse in Cherbourg. I wanted to get rid of the parachute. Still, I couldn't open the ripcord, preventing me from pulling out and inflating the dinghy. So I had only my Mae West, and the heavy parachute complicated my desperate situation. My body was lifted above the water while my head was pushed backward. I quickly became exhausted and was in a semi-conscious state.

After about 45 minutes, a French fishermen's boat discovered me. They first thought I was German but spotted RAF insignia when they cut off the ropes. They became so joyous that they started to kiss me and apologized that they had to hand me over to Germans. I threw out the pistol, gave them an emergency kit with money, and we went towards the coats where German patrol awaited me."

His broken hand was treated in the hospital in Cherbourg, and then he was sent from one German camp to another. Then, in

March 1943, he finally got to Sagan, where he met again with Ivo Tonder from the same squadron, who was shot down in the same operation the same day. Dvořák got heavily involved in The Great Escape preparation; he was a very handy tailor. He was given number 14 in The Great Escape for his effort, and his partner was Desmond Plunkett. They were supposed to go by train.

Here are his post-war recollections of that dramatic event:

"Once we left the tunnel, we went to Sagan train station. At 23:00, the air raid was announced, and within a few minutes, our fast train arrived. It was going to Wroclaw, so we got there without tickets and sat in the third class. We met Roger Bushell and his partner Scheidhauer there on 25. We arrived in Wroclaw at 12:45 only to be told that our train to Klodsko at 01:00 was canceled. So we waited in the lobby until 06:00 when we had another connection to Klodsko. During that time, we met Bushell, Scheidhauer, and Stevens again. We got to Klodsko at 11:00 changed the trains, and went by third class to Bad Reinertz. It was snowing, but we set on the walk to Nový Hradek in Czechoslovakia, where we stayed in the hotel until the evening of March 28, 1944.

Then we went to Nové Město nad Metují in Eastern Bohemia, where we hid in the barn on one farm until April 1, 1944, when we went to a village called Spy, where he hid again for one day and from train station Opocno, we went to Prague where we came in the evening of April 3, 1944.

We walked around Prague until 04:00 on April 4th and then took a train to Kolín, where we spent one day. Through Pardubice, we returned to Prague, where we wanted to meet a man who had helped John Stower in his last escape. This man owned a pub in Praha-Kbely. He generously allowed us to stay with him overnight, providing us with food and ration tickets. Despite the risk to himself, he was a courageous man.

We then went to Klatovy, where we were stopped by the Czechoslovak police, that took us to their station and held us there for two days. To our shock, they then handed us over to the Gestapo. I was disheartened as a Czech pilot who had escaped his occupied country to fight against the Germans, been shot down, escaped from

a German POW camp, and then was captured in his own country. This showed something about the character of the Czech people. The Gestapo took us to Prague, where we were held for five months until November 21, 1944."

Dvořák was then separated from Plunkett and was in Prague jail until November 30, while Plunkett was alone in Gestapo custody. Then, at the beginning of December, Dvořák was sent to Stalag Luft I Barth. He has stayed for about five weeks before being transported to Oflag VI C Colditz with Tonder and others. The American army liberated this camp on April 16, 1945.

Before returning home, Dvořák was promoted to Major and then has worked in Prague in the Air Division. In May 1947, he became the commander of the Air Base in Pardubice. A year later, he got fired from the army and sent to the reserves. He was then demoted to private and evicted from his flat. As a former "Westerner," he could not find employment and was forced to work the worst manual labor. He later worked as a clerk, but due to his poor health, he was granted an invalid pension at 49. He lived the rest of his life in obscurity. He died on an operating table in a local hospital due to a doctor's mistake on August 29, 1973, at 61.

(Sadly, not much is known about his post-war life. His brother, who still lived in the new millennium, refused to speak with us and provide us with more information to shed light on the dark spots in his life - V.F)

11 Chapter

JOSEF ČAPKA
One-Eyed Smiling Jo

Josef Čapka

Josef Čapka was born on February 23, 1915, in a small village called Kokory near Přerov. He studied at an electro-technical school and was trained as a pilot between 1936 and 1937.

He recalls the dark days after mobilization in 1938 and the German occupation:

"In May 1938, mobilization was announced, and I moved with my flight to the airfield in Krizanov. We spent our time there listening to the radio, which brought news that was not very positive. Next to the barn, our dining room was parked the "flying coffins," French planes Bloch 200, which I didn't like and doubted would withstand the Germans. After six weeks, the situation improved, and the mobilization was canceled. We returned to our base with mixed emotions, to the point where we were disappointed that the war wasn't on and that we couldn't show the Germans our determination.

A few weeks later, we were moving again, this time to Slovakia, because Hungary was showing signs of unrest and activity. But we didn't stay there for long because the second mobilization was

announced, and we moved back to the airfield in Brno. The bombs were loaded on the hooks under the wings, and we were waiting for orders to fly. But, instead there was a broadcast announcement about the annexation of the Sudetenland. After that, the room was silent. The atmosphere was tense, and we were told to expect further orders. So, in the meantime, we unloaded the bombs and started delivering mail because some important lines were cut.

The Christmas period was very stormy, and we heard the news that France was willing to sacrifice us to maintain peace in Europe. These rumors were confirmed when the Germans invaded Czechoslovakia on March 15, 1939. The radio broadcast this news and begged the public to remain quiet, "Don't resist. They sacrificed us to prevent the war."

Not everyone obeyed this order, and there were places where brave Czechs fought the Germans, but to no avail. The weather was terrible, windy, and snowing lightly from low clouds. We were sitting in the dining room when an airman ran in and told us that three German planes, Fiesler Storch, had landed on our airfield. These were observation planes of an unusual shape, and none of us had seen them before. The Germans probably didn't want to send anything more valuable before discovering if they would face any resistance. So, we got up and saw them for ourselves but didn't get very close to them.

To our horror, we discovered that they were protected by Czech guards, wearing swastika armbands. We were shocked by this treachery and stood there with our mouths open, feeling a stab in our hearts from our countrymen, who were now chatting friendly with the enemy pilots who stood by their planes. As we stood there, two formations of Me 109 dropped from the sky and made two circles above us, while Junkers Ju 52 landed and armed troops left their bellies. After that, the Me 109 landed too.

The Germans had everything perfectly organized as if they knew our airfield as the back of their hand. They positioned the planes at the places they had chosen beforehand and put bigger number of Czech guards with swastika armbands there. As the Germans passed us from all directions, none of them spoke a word to us. They completely ignored us. So, one of the traitors had to tell us to retreat back to the building, which is what we did, but we told him what we thought of him.

Finally, our commander called us all together and told us to be patient and not to do anything foolish. We had to wait for orders from our Ministry of Defense. Standing there, I realized this was the end of my flying in Czechoslovakia. I should leave before the Germans kicked us out or arrested us."

So, no wonder Josef didn't wait too long and left his country on June 19, 1939. After saying goodbye to his parents, he boarded the train to the border with Poland. On board the ship *Castelholm*, he got to France on July 30, 1939. He was trained in French fighter planes and had made several combat missions. After the fall of France, he reached England, where he landed on June 19, 1940. He joined the RAF and was first sent to 11.OTU in Bassingbourne before being sent to the newly-formed Czech bomber 311th Sqn. He had many close shaves and nerve-wracking experiences but still managed always have a smile on his face. He explained it after the war:

"Every time I climbed into the plane to fly above Germany, I was fearful. It was much worse since I couldn't show it to my crew. So, my defense mask was that even in the most dangerous situations, I put a smile on my face. Because of that, I earned the nickname "Smiling Joe". I only hoped that nobody saw through that my mask hid the big fear of the upcoming chance to die. My real and tangible fear was that we would die in flames or our plane would smash to the ground. My purple underwear and St. Maria cross always flew with me, but sadly they couldn't beat the fear. It always sneaked into the plane and sat next to me.

Some airmen didn't fight the fear as successfully as I did, and they started drinking. On the plane, he felt a big headache, was weak, and couldn't concentrate on his job. Drinking was accompanied by tiredness from flying, officially called L.M.F., the lack of moral fiber. We were all human beings, and the inability of the nervous system to withstand extreme pressure couldn't be called more callously. It wasn't difficult to figure out who had that problem. He

spoke loudly about flying, where he is the brave one and one shot of alcohol was followed quickly by another.

Sooner or later, such a person didn't come for briefing, reported himself as sick, or simply left the base when the operation was announced. I had a similar case with my bomb aimer, who always announced he was going to the bomb bay, but when I called him, he didn't reply. I thought he forgot to plug the intercom, so I sent a radio operator who found the bomb aimer lying unconscious. When it happened again, the penny dropped, and when it happened for the third time, I told him to report sick since he was endangering the lives of the whole crew.

The spring came, and the number of my missions climbed to 50. With each trip above enemy territory, I was scared and was always convinced that this trip would be my last, fatal one. Then I had an accident I never told anybody about, but I felt guilty.

We flew to bomb Kiel, but the clouds were so thick that we couldn't see a thing, and it would be silly to bomb under such conditions. The secondary target was Wilhelmshaven, so I asked the navigator to give me a course. But as it seemed, the clouds didn't get any thinner. Then, out of the blue, I spotted a hole in the clouds and a small city. There were no lights, and everything looked peaceful. Usually, I would keep flying and didn't take notice, but this time I felt in a good position, so I pressed the button and dropped one bomb without aiming. When I made a circle, I spotted that the hit was spot on. Even without aiming, the bomb dropped in the middle of the center of the town, and the buildings around just collapsed. After seeing that, I felt like the worst scumbag on Earth.

Why have I done that? I probably killed some fucking Germans, but these people were peacefully sleeping in their beds and probably around. There were no military targets. I felt immediate hate toward Germans, who invaded my country and forced me to leave it. I was irritated by the peaceful atmosphere of that city. Why should they sleep while I have to risk my life every night? All my replies were correct, but they couldn't excuse my actions. After that, I told myself I would return to that city after the war and find out what damages I had done, but sadly, it never happened.

When we landed and went for the briefing, I didn't say a word about that bomb, and nobody else mentioned that either. I wasn't sure if they protected me or even overlooked that. I knew that if I survived my next operation, I would have one more to complete the tour, and they would let me rest. My nerves were stretched to the limit during the last couple of operations. If we didn't reach the target in perfect position and I had to make another circle, I gave hard and long bollocking to the navigator since I felt closer to death with every rotation.

My 52nd operation was above Bremen. We dropped the bomb load, and while looking at the dashboard, I realized we had low oil pressure. I ordered the radio operator to pump up the oil, but despite that, it didn't go up. It looked terrible, and we got above the Northern Sea and flew at about 1,000 ft. We would have to land on the sea, but I wanted the crew to decide what they wanted. I told them the situation was terrible and we had a choice, land immediately on the sea or try and keep on flying. After a minute of silence, each said he wanted to keep flying. So, I did that, and with the last gallon of fuel, we made it to the base. I felt exhausted when I rolled to the position and switched off the engines. The whole crew left the plane, and I stayed in the cockpit. My legs refused to listen to the orders of my brain. I don't know how long I was there, but I heard the call of the squadron commander who came to congratulate me. I climbed on the grass, and he shook my hand:

"Well done, we thought this time you have had it."

"It was a doodle," I replied, turned over, and vomited.

I was so tired that I almost missed lunch the next day. I wanted to see the plane, so I went there and was welcomed by a mechanic.

"You are lucky to be alive," he said, motioning me towards the wing that looked like a strainer. The oil tubes had holes in various places, so we couldn't have increased the oil pressure with pumping. I climbed inside the plane and spotted shrapnel holes everywhere. One shrapnel was stuck at the seat of the radio operator second went through the navigator's place. They would have been dead now if they had both been at their places. Unfortunately, the right engine seized after landing, and the second wouldn't work

CHAPTER 11: JOSEF ČAPKA

much longer. I was convinced that my guarding angel was working overtime this night for me."

He was the first Czech member of the 311th Sqn who survived a whole tour of 200 hours above enemy territory and got a DFM medal. Then, after a much-deserved holiday, he became an instructor at O.T.U... He even participated in the first two "Thousand bombers air raids" above Germany. But being an instructor didn't mean having a life in safety. On the contrary, in February 1942, his pupil made a mistake he couldn't rectify. So, the whole crew had to bale out before the Wellington smashed on the ground and caught fire.

He soon got transferred to the newly established Czech/British night 68th Sqn equipped with Beaufighters. It was here when he experienced the very unusual but almost fatal incident on June 27, 1944.

"My observer was Willie Williams, and we got above the French coast when I spotted some dots in the blue sky. I reported that to the base by radio. When we got closer, we recognized Liberators, who must have been poorly done since they weren't flying in formation but were scattered all around. But they looked fine enough to make it back home. Then another dot appeared in the distance, and it was apparent that our flying lines would have to cross very soon. It was another Liberator but flew much lower than others.

When he got closer, I spotted he had silver paint, which wasn't unusual. But Willie told me something I had overlooked before. Liberator was flying with three engines and probably was severely damaged. There was a big hole in the plane's body, probably after the flak direct hit, and one wing was also covered with holes from shrapnel. But there was something strange about it that I overlooked at first sight. He was flying opposite direction. He would get back above the Channel and to France by this course.

We discussed it with Willie, and once we got very close to the plane, we both agreed that the plane was obviously in trouble and needed our help.

"Poor sod," said Willie, and that made me act. In my thoughts, I imagined a badly hurt American pilot wrestling with control. That

Josef Čapka on the right

decided the matter. I got behind his tail and wanted to fly past him and figure out how to help him. While passing him, I spotted in the corner of my eye, that his tail turret was following us, but it didn't look strange.

In one moment, I observed the limping Liberator… The next moment was complete darkness.

I didn't hear anything and even didn't feel any pain. I got instantly blind. It was as if someone had blown the candle. I didn't see anything, and my head vibrated as if it were part of the plane. I felt the cold air stream on my face, which should not be there, but other than that, all seemed normal. I instinctively pulled down the nose of the plane in fear we would collide with Liberator and then dropped a control stick by one hand and touched my face. An oxygen mask is gone, and instead of my left eye, there is just a sticky hole, no eye. I touched the right eye and felt utter pain. The next second I sat motionlessly and tried to figure out what had happened. Blindly I leveled the plane and carefully touched Willie's place expecting in horror just a pile of flesh, bowels, and blood. Thankfully, he was all right, so I punched him and shouted

"What a fuck had happened?"

"That fucking bastard… That fucking Yank gave us a burst. The left engine is dead."

CHAPTER 11: JOSEF ČAPKA

Now it dawned on me that we could catch fire, so I tried to find the button of the fire extinguisher, but Willie grabbed my hand and told me.

"I have done it already, no fire, we are OK."

So, we shouted at each other, and I was pleased he didn't see the other part of my damaged face.

I was relieved for a second that we were not on fire, but it quickly changed because I was blind, and we faced death.

"Willie, you will have to bale out. But don't worry; I will jump after you."

"No way, we are above the sea. Couldn't you try and make it above the coast," he replied.

We turned the plane in the homeward course with Willie's help with the control stick and because I was using him as my eye. Funnily I didn't feel anything, not even the pain, but without eyes, completely blind, I felt useless.

Thankfully, I let Willie fly the plane in the past, so he knew basic instructions.

"Take over, Willie," I asked him and was surprised when he refused.

"What's up, are you injured, or is the dashboard gone?"

Willie hesitated, and then he leaned towards me and shouted into my ear.

"I can't see a bloody thing. All are covered with blood."

I touched my face and felt the pouring of blood. Willie shouted into my ear that we were crossing the coast.

I pulled his sleeve and asked him if he would manage to land.

"No, not with one engine," his reply brought me into reality since I didn't realize that landing with Mosquito on one engine by an inexperienced pilot would be a suicidal attempt. To make matters worse, Willie reported that the trapdoor above the pilot's seat was jammed and couldn't be opened, every second lasted as a minute. Here I was above the English coast, blind, and I have to die. I have been freshly married to Rhoda. I haven't done anything wrong, so why must I die? Why me?

I got carried away with my thoughts that I had forgotten Willie.

I recalled another trapdoor under my legs when I returned to reality. Willie insists I go first. I tried. God knows I tried but to no avail. Once I let go of the control stick, the plane turned upside down, and I had to wrestle it back with Willie's help.

So, I persuaded Willie to jump first, and I will follow him. I didn't hear him go, but suddenly I felt lonely. I stretched my hand toward his seat, but it was empty.

He has gone.

I had no idea what height do we have. The last time Willie told me, we flew at 3,000 feet seemed long ago. I leveled the plane and tried again to escape through the trap door. The more I tried, the worse it went. I realized that I got trapped by a parachute harness. I had to use both of my hands to get myself free, but the plane again tended to turn on its back, so I had to get back to my seat and grab the control stick. As I got there, I felt something sweet pouring into my mouth. I know that feeling, and I know that I will pass out soon. I felt sick.

I thought it would be more merciful when I died unconscious. I won't know about that. My mum told me flying is dangerous and that all pilots will get killed.

"You were right, mum," I replied to her.

But then, suddenly, I felt the energy and will to live. I don't want to make Rhoda a widow after two weeks; it would be unfair; I must live.

I lifted my right hand and tried to open the lids on my right eye. Thousands of needles were sticking into my head, and the pain was cruel, but I saw land under me and spotted a dashboard. Willie was right. It was covered in blood. I stretched my right hand to clean the gauges but doing so, the eyelid got shut, and I am blind again.

I told myself not to hit my head and lose consciousness since the plane would likely catch fire, and I would have fried to death.

For a moment, I opened the eyelids with my right hand again and spotted small trees underneath me. Since all are of the same height, it would slow me down.

I pulled my knees to the chin and embraced my head. Doing so, I, of course, got blind again and didn't hold a control stick.

A mighty bang, and I heard a sound of breaking and tearing.

CHAPTER 11: JOSEF ČAPKA

When all stopped, I was on the floor of my cockpit. When I came to my senses, I realized I forgot to open the flaps and slow down the speed, so I must have hit the ground at about 200 mph. Another thump, and I felt real pain. My left arm, face, and forehead hurt like hell. I felt the earth under my feet and tried to run away. The parachute was clipped to the harness and hit the knees from behind. It made running almost impossible, so I felt like being in slow motion trying to get away from a blazing inferno but not fast enough. But the noise seemed stronger, so I probably ran the opposite way. Something hit me, and I fell onto the ground.

Then I turned on my back and tried to get my breath. A mighty explosion interrupted me, and I heard pieces of the plane flying all around me. It took me a considerable effort to force my body to crawl away from the heat. When it was all over, I removed my silk gloves and put them into the hole where my left eye once was. Doing so, I felt the piece of skin with hairs dangling from my face. That's the last thing I remembered."

Josef was taken to a hospital in Colchester, where much work would be done on him. He was visited by his wife, Rhoda, and the 68th Sqn. The Wing Commander Hayley-Bell wanted to know all details about his flight and landing. He also told him that squadron mechanic Jimmy Brown came to his office and seriously offered him one of his healthy eyes for Josef since the pilot needs two eyes. In contrast, a mechanic can do with one. Jimmy politely refused, but his gallantry touched Josef very profoundly.

After two weeks, sight in his right eye returned, and he could ask for a mirror to see his face. It was a bloody mess full of stitches, and he felt like Frankenstein. The left part of his face was covered with a bandage. He was transferred from a hospital in Colchester to a similar institution in Ely near Cambridge. The nurse brought a smaller version of the bath filled with salty water. His left arm was put there, and now Josef could see the full extension of his injury. The elbow was terribly burnt; he could see the bone through fried yellow flesh, and he was horrified when a nurse took the scissors and started cutting out the loose meat. The eye specialist did his best for Josef and informed him about Archibald

McIndoe, the plastic surgeon from East Grinstead who visits Ely Hospital every two weeks. When he sees a patient with burns requiring plastic surgery, he sends him to his hospital.

Rhoda visited him as often as possible, and they could walk outside the hospital. First trips to the town and meeting other people were a nightmare for Josef; he felt terribly low when people looked at him, and it was apparent they felt really sorry. But he didn't require their compassion; he wanted to be an average, healthy person and not be considered a protected species.

From the look of these people, he realized that his chances to keep on flying after the war were minimal.

On August 22, 1944, he became a patient of McIndoe in East Grinstead Hospital. When he came there, McIndoe told him about František Truhlář, who arrived there for the second time. They knew each other from the 311sq, although they never flew in the same crew. Josef received the bed next to Frankie's, and he also met and befriended Jimmy Wright, František's friend who was burnt and blind. Josef felt ashamed that he took so much self-pity when his mates looked much worse. Once, Josef was visited by an Intelligence officer who asked him about his injury:

"We went after that Liberator and realized he didn't belong to any base in England."

"Where was he from then?" asked Josef

"From Germany. That Liberator was reported as missing from the raid above Germany. It was shot down there, and Germans put it together so that it barely flew, and they used it for 'pirate' actions. We know they used this trick once before."

"How do you know that?"

"Your observer remembered all details he revealed to us at the briefing. They fit into other reports we had about it from other skirmishes. I know it sounds weird, but I want you to believe that since that is the truth," concluded the officer.

Just ten days before Christmas 1944, Josef received a left glass eye. Still, he wasn't very pleased with it since the eyelid was too swollen and uncomfortable. He was released from the hospital on March 15, 1945, and sent to the RAF depot in Uxbridge. He was

also informed that the glass eye is only provisional and will soon be replaced by a more modern one.

The war ended, and Josef survived. He was itchy to write home and hear any news from family too. But his injury needed more significant attention, and he had to return to East Grinstead. War was over, but treatment wasn't; McIndoe still had much to do. His Guinea Pigs needed him very much.

After the operation on June 7, 1946, he started to ponder his future and went to London to Czech H.Q. Much to his horror, he realized, it was empty; only NCO on duty advised him to go to Czech Repatriation Office. He was told he could only take home 25 kilos, and the cargo train had already left, so he must go with another one only for personnel. It didn't suit him, so he got a place in one Liberator which took off from Blackbush airport. Three planes were on their way to Prague, but only two had landed. Unfortunately, one Liberator crashed just after take-off, and all on board were killed. Josef again has luck on his side, even after the war.

He met his family and left for England to prepare Rhoda to return to her new home. They moved from Prague to Olomouc, where he became a commander of refreshing course. They visited the English club twice weekly and lived quite a happy family life. But the atmosphere in Czechoslovakia got more authoritarian, and after February 1948, it was apparent which way it was leading.

Soldiers who served in RAF were dismissed from the army. They were officially sent for a long-term holiday, but in other words, it was a sack. Josef Čapka was on top of the list. He went to Prague to the Ministry of Defense to prevent oppression. He asked for an emigration passport, which was promptly denied. While in Prague, he met an ex-WAAF who worked at the British Embassy and told him about the possibility of escaping from Czechoslovakia by foul line. He teamed up with Josef Bryks, another airman from RAF on top of the list, and discussed the escaping possibilities. They wanted their English wives to go out first, and then they would follow them. But before this could happen, both were arrested. He was accused of espionage and high treason, but the jury announced this didn't belong to their power.

Hence, they proclaimed it was a matter of airman's superiors.

This verdict didn't go down well with prosecutor Major Vaš who appealed against it and required a new trial. This time Josef got ten years in prison with heavy manual labor. Major Vaš, who considered everybody a foreign spy was finally satisfied.

During the first year of his imprisonment, he was in solitary confinement; one day in a month, he was left with no food; another day in the month, he was in total darkness; the next day in a month, they took away his mattress. To keep occupied, he got big boxes of needles and pins, which he had to sort out according to size, and big boxes of feathers, which he had to strip. Finally, after a year, he was sent to a giant cell with 30 other inmates. After five years, Josef applied for amnesty, which was denied since, as an officer, he conducted serious crimes and must serve the whole sentence.

After another 1,5 years in jail, new inmates were accepted. Josef spotted them and couldn't believe his eyes when he recognized his prosecutor, Major Vaš, among them.

He came to him, looked up his eyes, and asked:
"Do you recognize me?"
"No," replied Vaš, and they wanted to leave.
"Come on, not so quickly. When you don't recognize me, you surely remember the name of Air Marshal Janoušek, don't you?" asked Josef.
"I heard that name long ago but didn't remember," admitted Vaš, wanting to leave.
"You fucking swine," exploded Josef and stretched his arm for a punch.

Josef continues:
"In no time, he dropped to his knees and begged me to leave him alone; he was shivering in fear. I didn't hit him since I fully understood the irony of my destiny. Here kneels the one, who asked for my imprisonment, who wasn't satisfied with our first charge and demanded a new trial. Now he was begging me to be a prisoner himself."

The fact that I didn't hit him didn't mean that he escaped the punishment and vengeance of others. There were too many of those whom he destroyed their lives, so there was no chance he

Dennis Hickin, Josef Adam, Josef Čapka

would escape unharmed. He was beaten very often and had to report falling off the stairs; otherwise, he would have been hit more."

His wife Rhoda left for England, and he stayed in jail until 1954 when he was freed after the personal intervention of Czechoslovakian president Antonín Zápotocký. He even got an emigration passport and could leave for England to accompany his wife, Rhoda.

He didn't want to reveal details about this part of life. It remained a mystery since something like that was unusual in the 50s. Fifty years later, it was found out that Josef Čapka signed a cooperation with the Secret Police, and he was given the code name "Javor." This allowed him to be released from jail and move to England. However, it was only his trick to escape the prison and pretended to be on the other side. He was playing his game with them, and they were playing their game with him. He didn't reveal any information to anybody and lived his family life with Rhoda. The Secret Police terminated the cooperation as non-perspective. Josef Čapka worked as an electrician in a London power station. Still, war injury and prison left a heavy legacy on his health, and he died on July 27, 1973, from a massive heart attack. He was only 58 and buried at a small St. Peter Cemetery in Easton. Rhoda re-married his 68th Sqn mate Dennis Hickin.

12 Chapter

ARNOŠT VALENTA
One Who Didn't Return Home

Arnošt Valenta - Graduate

Arnošt Valenta was born on October 25, 1912, in Svébohov near Zábřeh in Moravia. After his father's death at the beginning of World War I, his mother married a neighbor, leaving her alone with three kids. Valenta attended secondary school and was a very clever student, but he longed to become an aviator. Unfortunately, his family's money for education was insufficient, so he enrolled in the national service and later applied to the Military Academy in Hranice na Moravě. Valenta's commander was Karel Klapálek, who became famous during World War II for defending Tobruk with a Czech unit against Rommel's army.

Arnošt graduated on August 1, 1936, and was sent to the 39th Intelligence Regiment in Slovakia. He enjoyed working with maps and military plans and intended to stay in this position. However, events in 1938 prevented his plans from coming to fruition. He then left for Poland, where he signed a contract to work as an intelligence officer for the Polish military staff and sent information about the location of the German army on the Protectorate soil.

In April 1939, Arnošt was accepted as a student at the Faculty of Arts at Charles University in Prague. During this time, he continued his secret mission, meeting various people from the military and political life. On August 21, 1939, he said goodbye to his hometown and left for Poland. He traveled with others to the former WWI military camp Malé Bronowice. Then, he was sent to Tarnow and Lviv, where he was shocked to hear that the German army had invaded Poland and the war was on.

Czech military units were organized, armed with five heavy machine guns, and the crew set off on a journey to Brest port. Stuka dive bombers attacked them, and there were the first casualties among the Czechs. They intended to head towards the border with Romania. Still, it was too late, as Russian troops had already invaded the eastern part of Poland, forcing the Poles to evacuate the territory. The Russians disarmed the Poles and the Czechs and allowed them to enter Ukraine. Arnošt admired the endless fields and saw the poverty-stricken villages, hungry women and children dressed in rags, and Stalin's portrait in every house.

The Czechs were accommodated in Kamenec Podolsky, but the conditions were appalling: bad sanitary conditions, poor food, a lack of water, and a lack of activity within the unit. They finally reached a half-damaged monastery in Jarmolinky. Arnošt wrote in his diary that the conditions worsened over time, and the political differences between individuals in the unit became sharper. Some attempted to desert, which led to a stricter regime from the Russians. Fights and theft were daily, and finding someone to manage over 600 people was difficult.

On March 9, 1940, they finally left with a bitter taste in their mouths, feeling disappointed and let down by the communist state. On April 14, 1940, Valenta and other would-be soldiers anchored in the French port of Marseilles. Valenta worked in a military office in Paris. After France gave up, he and about 150 airmen headed to Port Vendres. Under the escort of a British cruiser, their ship *Appapa*, the leader of a 13-strong convoy, set sail to Gibraltar and then to England. All Czechs received uniforms and began their training, but the process was not without difficulties.

Arnošt Valenta - second from left, Prague 1939

Valenta´s crew. - Arnošt Valenta 4th from the left

Arnošt was assigned to a radio operator's course, but it was poorly organized and ineffective.

The lack of English proficiency was a significant challenge. The navigators had to start from scratch since they were taught navigation through calculation in Czechoslovakia. The British method of astronavigation was utterly unknown to them. The crews were formed when their squadron was finally established and training was completed. However, there was still a lot of tension between the Czech and British staff. The pilots were held to higher standards, and the first few months of operational flying were exhausting due to a lack of rest. Unlike the British crews, who had a day off after an air raid, the Czechs were required to attend roll calls, marching exercises, and search for German paratroopers. There was a shortage of ground personnel, spare parts were scarce, and they faced hostility from the British. This escalated on October 16/17, 1940, when they lost three crews. Arnošt felt that serving in bomber command was a dangerous and suicidal job, and he grew to dislike it more and more. When they moved from Honington to East Wretham, their situation worsened, as their huts lacked central heating and running water.

On February 6, 1941, the Wellington 7842, with Arnošt on board, was returning from a successful mission over Germany when the radio stopped working. Emil Bušina, who had replaced the original navigator, Jaroslav Partyk, who became ill at the last minute, could not handle the situation and failed to give them the correct course. With almost empty fuel tanks, they had to find a place to land;

CHAPTER 12: ARNOŠT VALENTA

unfortunately, it was in France. The Germans captured not only the undamaged plane but also the entire crew.

The crew was transferred to Dulag Luft for interrogation and then to Stalag Luft 1 Barth before finally arriving at Stalag Luft III-Sagan. Arnošt gained a strong reputation among his fellow prisoners of war, not only because he spoke English, German, and some Russian but also because he could read German newspapers and compare their news with information obtained from a hidden radio. He also knew how to gain cooperation from the German guards and, through them, received valuable information about transportation in Germany, train departures, new regulations, and the layout of the camp.

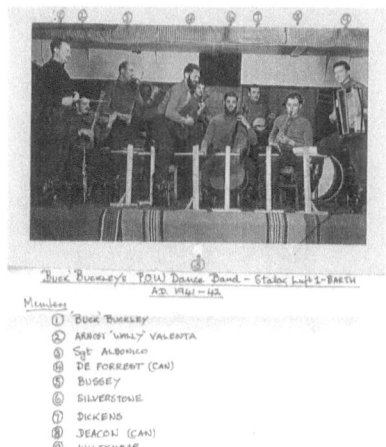
Arnošt Valenta Stalag Luft III - Sagan, A.V. 2nd from left

With the help of his tamed guards, he obtained wire cutters, ID cards, holiday and business trip forms, stamps, uniform buttons, and even a tiny camera, film, and film developer. Due to his valuable efforts, Roger Bushell chose him directly for the Great Escape. He forged documents for the name of a builder named Müller from Duisburg and was paired with Johnny Marshall. They were to leave Sagan by train from the local station.

However, due to an air raid, everything was in darkness, and Valenta and others could not find the entrance to the ticket office. With many essential trains already leaving, Marshall and Valenta agreed it would be better to continue on foot rather than take a train with multiple transfers. At dawn, they reached the Wroclaw highway, crossed it, and entered a deep forest, where they stayed for the whole day. After dusk, they were caught in a small village.

The Germans organized a massive manhunt, and the two escapees were trapped in the net. All police stations were provided with photos of the 76 POWs who escaped, and trains, buses, train

Stanislav Valenta - uncle of A.V. with author Svebohov, 2010

Author with the memoial to 100th anniversary of A.V. birth Svébohov, 2012

stations, hotels, cars, and villages were searched by the following evening on March 26. Valenta met 19 inmates in civic prison in Sagan, where the Gestapo started with interrogation. Then they were sent to jail in Görlitz and crammed four into small cells. Then the Gestapo loaded them into trucks, telling them they were returning to Sagan, but it wasn't to be. Germans completely breached the Geneva Convention and murdered 73 POWs who escaped from Sagan with a bullet to the back of their head.

Valenta was shot near the highway from Berlin to Wroclaw, and his body was cremated in Liegnitz. The official version was that they all were shot during an attempt to escape. Later on, fifty urns were taken to Sagan cemetery, and prisoners were allowed to build a memorial. Even the German guards showed the upset prisoners that they had nothing to do with the murders, and they distanced themselves from it.

(So Arnošt Valenta became the only Czech POW who didn't return home. His uncle Stanislav is still living in Svébohov these days. Although not well off, he raised a plaque commemorating his brave relative- V.F.)

Chapter 13

VÁCLAV PROCHÁZKA
Landing into The Sea

Václav Procházka in RAF

Václav Procházka was born on May 2, 1908, in Chrudim, Eastern Bohemia. He attended the School for youth pilots in Prostějov and served his national service in the aviation unit in Olomouc. After the German occupation, he fled to Poland. Eventually, he boarded the ship *Chrobry*, bound for France from the port of Gdynia. After the fall of France, Prochazka made his way to England, where he received the necessary training and was eventually sent to the 24th Sqn in Hendon and later to the Czech bombing 311th Sqn in Honington.

Procházka documented his memories of his flight that resulted in his capture and his time as a prisoner of war. On October 20, 1941, the Wellington KX-E took off to bomb Bremen. Still, due to technical difficulties, they had to drop bombs on the secondary target, Emden. The return journey was even more dramatic, as the plane suffered engine icing, forcing them to fly lower and lower to get rid of it. Ultimately, they were forced to land in the sea, but luckily, the plane did not sink:

"It was 10:25 and Bedřich (Valner) switched on our auxiliary radio and transmitted three dots, three dashes, three dots SOS. I called both gunners to get behind our seats. If only our right engine would recover, we could try to fly above sea level and reach within the range of our rescue patrols, and there would be hope that they would find us. And really, as if the engine listened to my wishful thinking, it started to cough, and the droning became regular again. I wanted to shout in joy since the Dutch coast appeared below us and the endless plain of the sea after that. Our height is about 700 meters. After half an hour of long gliding flight, we should be in friendly territory—only half an hour.

The whole crew was standing behind me, and only Bedřich was still in his position, asking for targeting. At the same time, I tried to prolong the flight as much as possible. Although the hurt engine started to work again, our good old Wellington started to shake strangely, as if it felt its end was near. We still descended, and Josef (Zvolensky) switched on the light so we could see the flashes of the restless sea below us. I spotted the high waves and froze solid since I had never landed in the ocean and had only heard about it in theory.

Suddenly, the plane jerked, lifted its nose, and flipped over to the right-hand side. The second pilot helped me regain stability, and we both looked in horror at the left wing. In place of the left engine, there was now only an empty black hole. I don't know what I did at that time. Still, I vividly remember the whole crew huddling in the corner, silently hugging each other like scared animals. Finally, we plummeted down, just above the water... then a mighty bang on my head...

I swallowed salt water and couldn't breathe. The entire crew was washed out of the fuselage by a wave as big as a house. I was pressed into my seat and tried to open the trap door above the cockpit, fearful of drowning. My soaked jacket and uniform weighed me down, and I desperately tried to open that door. I managed to stand on my seat and pulled my head out. Before I could look around, another giant wave went over me. I lifted my head, feeling an immense sense of relief to be able to breathe again.

CHAPTER 13: VÁCLAV PROCHÁZKA

Václav Procházka (right) as the best man of fellow Frantisek Trejtnar

Wellington swayed, and I could see the stars shining above me. All around us was water, with no sign of land in sight. I heard shouts behind the fuselage as my crew mates fought for their lives.

I saw Josef trying to escape, but something was holding him back. Perhaps the others were holding onto his legs. Suddenly, he inflated his Mae West prematurely and became too wide to get out. The others helped him relieve himself of the jacket, and he shot into the water. Before seeing if anyone else had followed him, another giant wave washed me back to the fuselage.

When I finally managed to get out, I saw the entire crew sitting on the fuselage and shouting joyfully to see me alive. As I joined them, we looked up at the sky and realized we needed to get into the dinghy, but it was nowhere to be seen. Just as we were trying to figure out what to do, Josef jumped into the water towards the right engine and retrieved the dinghy, which appeared on the surface in no time. However, the current was carrying the dinghy away from us because someone from the ground crew hadn't

secured it to the fuselage. Brave Josef started swimming after the boat, jumped in, and disappeared into the darkness.

We were all in amazement and stared in the direction where we had last seen the shadow of the boat with our friend. I felt a sense of responsibility for the entire crew, including Josef. Still, I realized that in a few minutes, we could be worse off than him if the plane sank into the water and we all drowned.

Franta (Petr) devised an excellent idea and cut the canvas at the rudder, securing himself to the plane's structure with his hands. Suddenly, I thought we could contact Josef using signal rockets. The pistol was inside the fuselage, above the navigator's cabin. Even though I couldn't see anything due to the complete darkness, I dived into the water and found the box with signal rockets and a pistol. The package was wet, but I loaded a rocket into the gun and fired it. I fired three times, and Josef was firing his rockets back. However, after all our missiles were gone, Josef didn't come any closer but seemed to be moving away. Our last hope vanished, and Josef was gone.

It was strange that the plane didn't sink; it was floating. Finally, the sea calmed, and I could finally talk to my crew mates. We were about 10-12 miles from the shore, in the shallow waters near the Friesian Islands. I stood on the pilot's seat in the water, up to my waist. The clouds, that had caused icing and were our downfall, had disappeared, leaving us with a clear, bright night. The water was shallow enough that we had actually landed on a sand dune. Our friends were flying overhead, their engines droning like music to us. I imagined our crew being among them. Although we were sure we were sitting on the sand, we didn't dare move for fear of destabilizing the plane.

It was dawn, and the surface glimmered. I had knocked out six teeth during the impact with the dashboard, and my mouth hurt. I looked at my crew's faces and couldn't recognize them. The previous night had left an indelible mark on their faces that would stay with them for the rest of their lives. Bedřich was exhausted, and his face had shrunk. Erazim's fatigue was more than he could admit, and his forced grin couldn't hide it. But František (Petr) and Josef (Sůsa) still had youthful energy despite growing old.

CHAPTER 13: VÁCLAV PROCHÁZKA

Václav Procházka - France 1940, top row second from right

At around 09:00, a patrolling Catalina appeared on the horizon. As we no longer had any signal rockets, we stood up and frantically waved, but to no avail. The crew didn't see us, and we started to speculate if the plane was sent to search for us or if it just happened to be in the area. Our high hopes dropped to complete hopelessness. Finally, I had to give the most challenging order a captain could give: we had to start dismantling our own plane. Silently, we divided our tasks and began to disassemble the almost intact equipment of the Wellington.

Suddenly, Josef, disassembling the machine guns in the rear turret, screamed, "Look there!" and pointed eastward. He might have spotted a rescue boat that would take us home under the Germans' noses. But I was wrong. Thanks to the low tide, we could see a propeller sticking about 200 meters from us, like a cross next to our former left engine.

All morning, we alternated, looking around and trying to spot a glimpse of hope. But all that appeared was a plane at around midday.

It was a German plane, and he spotted us. We all sat on the fuselage as he made a circle above us. Then he flew away, much to our relief. Still, within 90 minutes, a tiny dot appeared on the horizon and grew bigger and bigger… it was a German rescue boat!

Although this meant rescue from the sea, it also meant the end of the war for us, and imprisonment in the hands of the Nazis. When the boat arrived about five meters from us, the five airmen of the RAF stood on the plane facing seven Nazi seamen. We looked at each other with curiosity rather than hate. They grinned at us and showed us they would help us with a rope, but the rope in the hands of two Nazis looked like an ugly symbol.

When they pulled us onto the boat, they took us below deck and gave us five overalls so we could change out of our wet uniforms and let them dry. Then they brought us coffee and offered us cigarettes. This friendly gesture surprised us and improved our mood, as well as our determination.

"What if we get into them and turn this damn thing towards the English coast? It would be 7 to 5," Frantisek (Petr) suggested.

Thankfully, Erazim (Veselý) was more level-headed than I was. "Are you fucking crazy? Do you think we could handle such a crazy thing? If we did, the first German plane would sink us!"

Upon arrival on shore, they took us to an old building, and we stood by the wall. They gave us our dried uniforms. We awaited interrogation, but the door opened, and Josef came in instead. He looked at us with surprise and was about to scream in Slovak, but our warning glances made him change his mind, and he silently stood with us.

Soon, two Germans in long leather coats entered.

"Who among you is Czech?" they asked.

We shrugged our shoulders and pretended not to understand.

"Nobody? Okay, as you wish."

One man with his hands behind his back lifted them, and I noticed my silver cigarette case. I realized that they must have searched our pockets, and without thinking, I took two steps forward and reached for the case. He quickly pulled it back, and I saw a picture of my wife in his left hand.

CHAPTER 13: VÁCLAV PROCHÁZKA

"Shit, I am done," I thought immediately. During all 25 air raids, I took the cigarette case as my amulet. Still, it never occurred to me that it could hold proof of my identity, like a personal ID card. Instead, the photo had a dedication in Czech on it.

He handed the photo to the second man and asked, "So you are from the Protectorate and fight against the Reich? You are damn traitors and will pay the highest price for that."

"I am originally from Czechoslovakia, but now I am a member of the RAF and under the protection of Great Britain. As such, I fought against Germany. I was never a citizen of the Protectorate," I said mechanically, repeating the rehearsed sentence.

"You are all traitors, and you will all be executed. You will go first, and tomorrow morning, I will personally shoot you," the man in the leather coat shouted at me.

Two more officers entered the room, which was now full, and they looked at us as if we were exotic animals. Finally, one spoke fluent English and said, "In England, there are imprisoned German airmen, and we will treat you as you treat them. Do you understand?"

We nodded in agreement.

The second officer studied Josef's face for a long time, and I was worried that the hot-blooded Slovak Josef would jump at his throat. Finally, the officer asked, "Are you also British?"

"Yes," Josef replied.

The officer laughed in perfect Slovak, "Your face can't hide your Slovakian origin, even from God or the Devil."

Both officers laughed at this, and as Josef later admitted, he was close to hitting the man's face.

After that, the two Gestapo men and two officers left. We were moved to a truck accompanied by a Luftwaffe officer. He told us, that the journey would be long but didn't tell us where we were going. After about two hours, we arrived at an airbase and were sent to solitary confinement.

The following day, we were awoken and sent to take a real hot bath. I had never relished a tub like that before. After a coffee substitute, I felt much better physically and mentally. I was so

Václav Procházka France V.P. in the middle

hungry that I grabbed a piece of dry bread that tasted like cake that morning.

After a short interrogation that lasted only a few minutes, we were pleasantly surprised when we were sent to a bus where ten New Zealanders and Canadians were already waiting. We became part of the Stalag Luft VIIIB Lamsdorf contingent. We were somewhat shocked by the number of 40,000 prisoners from around the world for whom the war was over. Some of them had been there since 1939, and they had no trouble sleeping anywhere and eating any food given to them. The Germans tried to respect the Geneva Convention, so we were mostly okay. Still, we discovered their inhumanity while wandering around the camp.

We had nothing to do, so we wanted to see the camp. When we came to a double barbed wire fence with a notice that said the guards would shoot without warning, we wondered what was behind it. However, we soon saw a 100-meter-long, 10-meter-wide, and half-meter-deep mud flat filled with moving creatures, surrounded by another double barbed wire. The creatures were Russian prisoners suffering and dying in the rain, frost, and half rations of food.

When we first saw this, we couldn't believe our eyes. Is this really happening? We were not prepared for such a sight. As we saw the roll call of the prisoners, guards shouted at them with vicious dogs, some of whom carried long rods. We soon discovered the purpose of these rods upon their return to the camp. The dead bodies of prisoners were tied to the rods by their hands and feet.

We remained in Lamsdorf until mid-April 1942, when we were sent to Stalag Luft III-Sagan. This was when our crew was split for

the first time. Life here was much more tolerable, and we regularly received Red Cross parcels, which helped to improve the monotonous camp menu. However, sometimes the entire train of food parcels was sent "by mistake" to German soldiers, and we felt the impact of this mistake almost immediately.

After the Great Escape, František (Petr) and Josef (Sůsa) were sent to Stalag Luft 1 Barth. In contrast, Josef (Zvolenský) and Bedřich (Valner) were sent to Stalag XI B. Erazim (Veselý) was the only officer from our crew, who stayed in Sagan. We thought we had reached the war's end, but things worsened when the Gestapo started interfering in camp life.

One day, a guard told me to go to the office of the camp leader. I complied and did not suspect anything terrible, only to find nine other Czech airmen there. Two men in black coats told us we would go to Prague with them.

Once we arrived in Prague, the interrogations followed a similar pattern. They repeatedly asked who had helped us cross the border and whether we would confess to high treason. After two weeks, there was another change of direction, this time for the better. They put all the Czechs together, transferred us to a prison in Loreta near Prague Castle, and all interrogations ceased. Within two weeks, we returned to Barth, receiving a hero's welcome.

We returned to our old quarters, and everything seemed to be back to normal. For about ten days, we enjoyed the "freedom" of the prisoner-of-war camp before another blow arrived. We were sent to Oflag IV C Colditz, where we faced the possibility of execution for high treason. Every morning, we woke up fearing we might not make it through the day. It was a nerve-wracking race against time".

This is where Václav Procházka ended his memoirs. He won the race against time when, on April 16, 1945, Colditz was liberated by the American army. After that, he was transferred by plane to England and, after recuperating, attended a refreshment pilot's course in Wolverhampton. After returning home, he worked as a flying instructor at the airbase in Hradec Králové. But obviously, 1948 put an end to his career. In May 1949, he was thrown out

Václav Procházka being decorated after the war

of the army. As if it wasn't enough, he was moved from his flat to Jablonec nad Nisou in the North of Bohemia. He found a job in a local factory for imitation jewelry. However, the Secret Police still pursued him, and he was thrown out again from his position and his flat. Finally, he went to Pardubice in Eastern Bohemia, where he got a job as a manual worker. Obviously, bad treatment during the war took its toll. He died ill and was totally forgotten on September 5, 1973, only 65 years old.

Chapter 14

PAVEL SVOBODA
Swapped Identity

Vernon Bastable -Svoboda's escaping mate

Pavel Svoboda was born on June 28, 1918, in Bohuslavice near Kyjov in Moravia. He wanted to be a lawyer and attended law school. When the Germans shut down the universities, they arrested many students. One of them was Pavel. He was sent to Sachsenhausen Concentration camp. He remembered that time:

"We laid next to each other, and there was no central light, so if one needed to go to the toilet during the night, it was a problem. Firstly, he had to be careful not trod on laying inmates, which was an almost impossible task. Then after nature's answering call and he returned, his place was already occupied."

After being released, he didn't wait long and left his home to Slovakia and across Hungary, Yugoslavia, Greece, Turkey, and Syria to France and from there to England. Here he joined the RAF and was sent to gunners' school at Dumfries. After that, he joined the newly formed Czech bombing 311th Sqn. The fateful air raid came four days after Christmas on December 28, 1941. His Wellington

KX-B bombed Wilhelmshaven, and they had to land in the Northern Sea on the return journey. They stayed six long days in the dinghy, exhausted and washed up on Dutch cost. Two of his crew members didn't survive the atrocities. Another one stayed in the rear turret and was buried in the plane at sea.

After recuperation, which took almost half a year, he was sent to Dulag Luft for interrogation and then forwarded to Stalag VIIIB Lamsdorf. He managed to escape from there but was caught soon afterward and got three weeks in solitary confinement on bread and water. This "healthy diet" didn't prevent him from escaping more times, but he was always caught and beaten up. However, the Germans didn't manage to break his spirit, and his escapes became increasingly sophisticated. His most successful attempt came in May 1944.

Pavel discovered that the best way to escape is swapping identity with a working POW and escaping from outside the camp. So he found a New Zealander who loosely looked like him and was called Harry Colin Neville. He moved instead of Pavel into RAF quarters, while Pavel in green uniform found a bed in New Zealander's hut. They both memorized each other's data which Germans had in their files and tested each other. Then they shook hands and promised that they wouldn't tell anybody anything. Pavel only informed about this to hut leader and representative of the Escape Committee. Newcomer couldn't be suspicious since fellow prisoners wouldn't trust him and helped him with an escape attempt.

With a strange English accent, "New" Harry Colin Neville volunteered for manual labor outside the camp. With twelve other English prisoners, he left for Malé Heřmánky to work in Hoffman's Sawmill. Not far from here flowed river Odra full of big stones and with lonely houses around. Pavel's group teamed up with another group of Australians who had worked here already for a few days.

The task they had to do was pretty easy, fold wooden logs and cut the planks to let them dry. They worked the whole day in the fresh air, which helped boost their physical conditions. Pavel took plenty of English cigarettes with him, swapping them for fabric paint with one German guard. He wanted to paint his green

uniform black to avoid being suspicious during his escape. He tried to escape alone, run only at night, and stay in the woods during the day. He knew the area not far from here very well. He was heading toward Dolní Lipová since he assumed he would reach Protectorate borders more quickly. But shit hit the fan when the Germans discovered he had disappeared. They alarmed everybody in the area, and although Pavel tried to avoid the roads and stayed in the wood, he was caught and was "free" only for 36 hours.

He was first taken to a sawmill and then transported to prison by Wehrmacht guards. He spent two weeks in jail and another two weeks in Lamsdorf while the actual holder of the name "Neville" enjoyed the privileges of being a prisoner of the Royal Air Force in a nearby camp. Pavel was held in a cell with only water and bread for another two weeks before being released and swapping identities with Neville.

After the Second Front opened in Europe, Pavel decided to try escaping again. While in the camp, he met Canadian airman Vernon Bastable from Winnipeg, who had been shot down after an air raid above Saarbrücken. After Bistable's unsuccessful escape attempt, he was jailed for 28 days and happily accepted Pavel's proposal to escape together. But instead, the pair decided to run to Bohuslavice, Pavel's hometown, and swapped identities with a New Zealander friend who was missing the comfortable life in the RAF compound.

The two then volunteered to work in a manual labor transport to a quarry in Malé Heřmánky, where a group of British prisoners was already working. After a few weeks, Pavel asked the quarry gunner for help, and they successfully escaped to a partisan group in Chřiby called "Carbon", which was commanded by František Bogataj. Pavel and Bastable lived in the woods with the group for four months, participating in sabotage operations and supplying weapons to partisans in Southern Bohemia.

On April 14, 1945, the Germans conducted an anti-partisan raid on a hut called Kameňák, where their weapons were exchanged. The partisans were shot immediately, but Pavel, wearing an RAF uniform, was only arrested. He was taken to the ill-fated Kounice college in Brun, where the Gestapo shot many people and

Pavel Svoboda - sitting 2nd from left

where he had already been jailed before the war during the Germans' crackdown on university students.

After nine days, he escaped with prisoners waiting for release and joined another partisan group, "Rada 3", led by General Luža. The group conducted marauding activities, making life difficult for the Wehrmacht in the last month of the war. Pavel continued to pretend to be a New Zealander. He spoke English, which the other Czechs accepted except for one person—Professor Dr. Josef Grňa-Vlk. Dr. Grňa-Vlk was an accomplished linguist and one of the leading figures in the resistance movement in the area.

He spoke with Pavel in English and discovered some grammatical differences in the "Kiwi's" spoken word, which made him suspicious. Many times before, the Germans had sent their man to infiltrate a partisan or resistance group, so he was already wary. He informed the partisan group commander, who ordered Pavel to come and question him with many uneasy questions. Ultimately, Pavel has to admit the truth and reveal his real identity, along with all his war activities, life as a POW, and escapes. His identity was confirmed the next day, when he met some local people who remembered him from pre-war days.

On May 9, 1945, Pavel applied to the Russian Embassy in Brno and was sent to Prague, Pilsen, Brussels, and Britain. Due to his ill-treatment in the prisoner camps, he needed rehabilitation for stomach problems. He returned home in September, not August like most of his comrades. He tried to find accommodation for his Danish wife Ellen and their son both still being abroad. Still, it wasn't easy, so he returned to England and organized a means of transport to Denmark for them. He returned to Czechoslovakia and asked the Ministry of Defence for a transfer to Brno. He then returned to Britain to bring his property back to his homeland.

He intended to fly with Jaroslav Kudláček's Liberator but changed his mind at the last moment, which saved his life. Unfortunately, Kudláček's Liberator crashed during take-off in Blackbush, killing all on board. Ellen and their son had difficulties returning from Denmark to Czechoslovakia but arrived in November 1945. However, post-war life in liberated Czechoslovakia was not happy for Pavel Svoboda.

He encountered many officers of the Czechoslovakian army, who served at home during the war. They obeyed the orders of Nazi officers without facing air raids, trenches, and other war atrocities. After the war, they immediately applied to join the Czech Communist Party, which pardoned all their war sins. The party found another devoted and obedient servant who followed orders, and anyone who showed inclinations to breach them was threatened with their war collaboration being publicly revealed. These officers were now in charge of the army, which tried to eliminate "Westerners" with democratic views and war experience and replace them with pro-communist lackeys.

On June 1, 1948, Pavel was stripped of his position in Brno, which triggered another escape from his home country before the communists could arrest him. The conditions and treatment in Czech communist camps were much worse than those in German POW camps. He sent his wife and son first, then escaped to Austria and Denmark. They eventually moved to England, where they lived until his death on January 8, 1993. In 1991, he was fully rehabilitated and met old friends, he hadn't seen in ages in Prague during the summer.

15 Chapter

OTAKAR ČERNÝ
Over The Barbed Wire in Two Regimes

Otakar Černý

Otakar Černý was born on November 28, 1918, in Křenovice near Vyškov in Moravia. In 1937, he graduated from grammar school in Brno and entered the Flying school for officers in Prostejov. He was later sent to the Military Academy in Hranice na Moravě. After the establishment of the Protectorate, Černý escaped to Poland and then to France on the ship *Chrobry*. After France's surrender, he went to England and was sent to the 311th Sqn as a wireless operator.

As part of the crew of Jaroslav Nyč, Karel Šťastný, Jaroslav Zafouk, František Knap, and Jiří Mareš, Černý took part in the 18th trip over enemy territory in his Wellington KX-N. The target was Hamburg, and the secondary target was Kiel. The visibility was abysmal, but the crew dropped their bombs and were returning home when a massive explosion occurred under the pilot's seat. All lights went out, the radio was destroyed, and the navigator was covered in oil. The crew was forced to bail out, but unfortunately, the rear gunner drowned in

the Zuider Zee. The reason for the explosion was never cleared, but all crew members believed it was sabotage.

Černý had a clear view of the events that took place. He had heard from the mechanics, that they had spotted a suspicious Czech electrician and an Irish mechanic near the planes. In addition, some ground crew personnel were Irish, didn't like serving Great Britain, and didn't do their job correctly. Černý believed the delay before their flight due to a light malfunction in the bomb bay resulted from sabotage.

After landing in the pitch dark, Černý hid his parachute under a hedge and heard heavy footsteps approaching. He thought a platoon of Germans surrounded him until the moon emerged from the clouds and revealed a herd of cows. He followed a ditch that led to a local canal and found a fisherman's hut. He entered and commandeered a blue pullover to cover his RAF uniform. He walked to the first town he came to, Sneek near Leeuwarden, and found a market where he got some food:

"Unfortunately, I only had a chocolate bar with me, which didn't seem like a lot for what I thought would be a long journey. I stayed there for two days, sleeping in a hut and exploring the embankment of Zuider Zee. A tiny road separated the lake from the sea, and I spotted some fishing boats that would have been an excellent way to escape. However, they were too big and heavy for just one person. Despite this setback, I didn't give up and continued to find a way out. Finally, my persistence paid off when I spotted a much smaller boat transporting fishermen to their larger ships. This became my escape plan. I decided to steal the small boat during the night, cross the road, put it in the sea, and paddle towards England. In the afternoon, I returned to the market to gather vegetables and fruit for my journey.

While I was there, two Germans on bikes rode by with rifles on their backs. I greeted them in my school German, and they replied. But for some unknown reason, they returned and asked for my documents, which I didn't have. As a result, I was arrested and had been taken to a prison near Leeuwarden airport. When I asked the Germans later why they returned, they told me that I was the

only person on foot, while all the Dutch people used bikes, making me suspicious. The next day I was taken to Amsterdam prison and to Dulag Luft for interrogation."

Then his final destination was Oflag VIB Warburg while his crew was dispatched to Lübeck and Bad Sulza, respectively.

Warburg camp was primarily for British infantry officers, and a small section of RAF personnel was also accommodated there. One of the most prominent prisoners was the famous legless fighter pilot Douglas Bader. It was tough to get used to the new life situation for Otakar. A few days previously, he was with his friends in the pub and with his new wife, and now he is behind wires on enemy territory. The prospect of staying there for a long while looked very tangible.

One of the most popular and common ways how to bide time was escaping activity. Otakar took an active part in it, and during the night on April 19, 1942, he escaped through the tunnel among the first five prisoners. He was waiting for his partner Canadian F/O Asseline, but he was nowhere to be seen, and instead of him came Josef Bryks.

Since their respective partners didn't show up, they decided to join forces and continue together toward the Swiss border. They continued only during the night and avoided villages and towns. Finally, on the sixth night, they appeared in the valley in front of the village, which had steep hills on both sides. They decided to risk it and went through the village, hiding behind each house. But somebody must have seen them and called the guards. Josef Bryks managed to escape, but Otakar Černý was caught at midnight and taken to the local prison. He showed the commander his forged documents and insisted that he was a Hungarian worker returning from the shipbuilding yard in Karlsruhe. The commander "bought" his story and told him he would be released the next day. But he didn't want to wait and tried to escape from the cell by sneaking through bars bare naked. But his attempt was aborted by barking dogs under the windows, which alerted the guards who caught Otakar, and that was the end of his escape. Next day, during the interrogation, it was found out, that the documents were

forged, and he was accused of espionage, which could have fatal consequences for him. The Germans brought some French infantry prisoners who escaped a few days ago and were very violent towards them. When Otakar saw how they kicked and beat them, he showed his dog tag and shouted that he was an RAF airman and would report this behavior to the Swiss Red Cross. It brought immediate effect, and the Germans stopped their aggression.

Otakar was sent to Stalag Luft III, where he met fellow Czech airmen. He stayed here for almost twelve months before being sent to Oflag XXIB Szubin. Soon after arriving here, he got involved in digging another tunnel finished in March 1943 and led from latrines to nearby potato fields. The escaped took place on March 3, 1943. Following days Josef Bryks and his partner S/Ldr Morris were taken out from the same camp by a "shit cart", hidden in the cask with rubbish. The escapees met in a big barn with plenty of hay, where members of the Polish resistance brought them some food.

Since Morris got sick and couldn't hold Bryks, both Černý, and Bryks got to Warsaw and connected with the local resistance movement. For almost a month, they worked as chimney sweepers in Bogumilski. But their luck ran out within three months, and the Gestapo captured both. It was revealed that one member of the resistance movement worked for the Germans and betrayed both airmen. Otakar Černý was kept in prison Pawiak where he underwent many interrogations. When it was revealed he was an escaped RAF Kriegie, the interrogation stopped, and he was left alone. He could send cards about his well-being; one was addressed to SBO in Sagan, where he asked for help. It worked, and after 63 days, he was sent back to Stalag Luft III along with Josef Bryks, who was badly treated and beaten by SS men and had to be operated on in a camp hospital.

In the camp, Otakar offered his service to the digging team and worked on tunnels Dick and Harry and the latter was used for The Great Escape. As a result, Otakar got on the list of 200 prisoners who were to escape by this tunnel. Still, since his number was well over 100, he never got out, which in hindsight, was a good thing

Otakar Černý second from right in POW camp

since very likely he would have been shot.

In the autumn of 1944, he was transferred to Prague and interrogated in Petchka Palace. He was sentenced to death for high treason and sent, as an intrepid escaper, to Oflag IV C Colditz, a camp for "prominent" prisoners. The first American soldier came to Colditz on April 16, 1945. When he stood at the gate, happy Czech prisoners shouted at him in their native tongue:

"Where have you been for so long?"

To which he replied in flawless Czech:

"Sorry guys, we have done what we could."

Then he introduced himself:

"Joe Štefan, Nebraska."

The world is small. American troops liberated the Allied POW camp full of Czech prisoners. The first soldier at the gate is an American soldier of Czech origin. It sounds like a Hollywood film cliché, but it is true.

The war ended, and Czech prisoners were sent back to England. All of them carried some memories from life behind the wire. Some of them mental, some of them physical. Otakar admitted that even

ten years after the war, he had wild dreams and nightmares:
"I had dreams about being chased by SS troops, and when they were about to get me, I woke up with a quick heartbeat."

He returned to Czechoslovakia and tried to get back to everyday life. In 1947 he went to England for a business trip as an interpreter with the Ministry of Defence officers. The task was to make a deal with the famous Rolls Royce about the production of jet engines and the possibility of producing them in the Prague factory, Letňany. The good deal was, of course, halted after February 1948. Otakar lived in a flat with his English wife, Rhoda, who was expecting a baby. So far, requests for a better apartment have fallen on deaf ears.

On May 6, 1948, the Chief military prosecutor issued an arrest order on Otakar Černý and fellow airmen Josef Hanuš, Jan Plášil, Václav Bozděch, Josef Bryks, and Josef Čapka. On September 1, 1948, he was released from the army with no back pay. His CV was written with a horrible evaluation, saying he has a very negative attitude to people's democratic regime. Similar words appeared in almost every CV of former RAF airmen.

After the war, 444 "Western" Czech airmen stayed in the army. Only 13 of them were left there by the late '40s. The remaining 431 were sacked, humiliated, and most were arrested. Otakar stayed in the famous Prague prison "Domeček" till January 6, 1949. The only consolation at that time was the news that his wife and son had managed to fly away from Prague to London with the help of the British Embassy. Nevertheless, Otakar was charged with three years of hard manual labor and losing all civilian rights and all war medals. He also had to pay all trial costs.

The intrepid war escaper didn't forget any of his skills. This time, he was escaping from a concentration camp in his own country, where the Czech people sent him to fight against the Nazis. Not long after his imprisonment, he started to build a plan to escape for one last time, hopefully to freedom.

He spotted that his inmates were reluctant to escape since they would put their families into misery. How could they live freely abroad when their loved ones would suffer under communism

Author with Otakar Černý

and have no chance to see them again?

He found two inmates, Josef Gapa and Vladimír Pechek, who were single and had nothing to lose and agreed to escape, too. They all knew what would expect of them in case of failure, more beating, more hunger, and more years spent in jail on top of previous sentences. The billet they lived in was in the middle row and had windows atop the roof. They put one table on top of the other and reached the windows protected by barbed wire, but they cut it with stolen pliers. They took off the windows and waited for the right moment when the guard passed under the billet and continued his walk. Then they climbed on the roof and jumped into the garden under the hut. As quietly as possible, they passed through the unused part of the camp, and as luck would have it, they found a ladder that came very handy when climbing over double barbed wire.

The trio went towards the border with East Germany, not the borders with West Germany as predicted by the Secret Police. When the alarm was set the next day and the search began, the escapees were in the mountains close to the border. There were many hedges and bushes where they could sleep during the day. They could see armed soldiers underneath them, but luckily soldiers didn't climb up. When the darkness descended, they continued their journey to the Southwest toward the American occupation

zone in Germany. The problem was a lack of food since they only had two loaves of bread between them, so after ten days on the run, they had to risk it and beg for food in lonely houses.

Crossing the border to East Germany wasn't a big problem since it was lightly guarded. They only had to crawl under the watch towers through the dry river bed. Once in the American zone, they reported themselves to the guards. Daily Express wrote about their successful attempt. While Otakar Černý was sent to Nüremberg, his compatriots were sent to an immigrant camp. They never saw each other again. American intelligence service CIC interrogated Otakar, and he told them all he knew. After two weeks on the run, he finally got delicious food. After about two weeks, when all information was filled in, he was sent to the camp near the Swiss border but instead of staying there, he told Americans about his intention to go back to RAF.

Life in the camp was impoverished, and there was a danger of Russian agents who could quickly get to and from the camp as they pleased. It was guarded by some Ukrainians and Germans, who didn't care much about their duty. Russian agents could kidnap anybody they wished and bring him back to Czechoslovakia. When Otakar heard they might be after him the following night, he slept in a different hut.

Luckily after a month of living in fear, he was finally handed to Brits on the airbase near Nüremberg, where he repeated his request to return to the RAF.

Further interrogation followed, and Brits prepared the conditions under which he could finally rejoin RAF.

When DC 10 brought weapons, food, and ammunition to the occupation zone, they took refugees as the return cargo. Later on, in Czechoslovakia, the property of Otakar was confiscated, and he lost almost 25,000 crowns. But it didn't bother him at that time. He was back in England and prepared to start a new life. In 1952 he received British citizenship, and two years later, he was dispatched to Edinburgh with Communication Squadron. He lived happily in England for 60 years and died as the last Czech POW.

16 Chapter

FRANTIŠEK BURDA
Flying Violin Player in Sagan

František Burda

František Burda was born on January 21, 1915, in Osičky, a small village near Pardubice in Eastern Bohemia. After graduation, he worked as a building engineer in Prague. In 1935, he went to the army for his national service and wanted to go to artillery, but he was refused since his weight was only 54 kilos. So, he applied to the school for youth pilots in Prostějov and was among 60 lucky chosen from 600 interested boys. He didn't know then that many of his schoolmates would be his brothers in arms in the 310th squadron within five years. A year later, his father died, and František decided to become a professional soldier. Unfortunately, the economic crisis also hit the building industry, so he had trouble obtaining a job after his return from the army.

In contrast, army life offered an excellent future to a young man. He became a military academy student in Hranice na Moravě. When he left it, he was one of the youngest officers in the Czech Air Force at 22.

When Hitler invaded his home country, he left for Poland, and on July 25, 1939, from port Gdynia, he cruised to France on board the ship *Castelholm*. Here he reluctantly joined Foreign Legion, and when the Germans invaded France, he was released to French Air Force. First, however, he had to undergo a long flying course to get used to French planes.

His first combat mission happened on June 3, 1940, and two days later, he met his first enemy plane and got entangled in a dogfight. Although he shot down one Messerschmitt Bf 109, it crash-landed in the smoke on the meadow near Amiens. After that, however, the whole squadron shot down seven out of ten enemy planes, and Burda could put one to his tally.

On June 6, 1940, he was called for another combat mission; this time, it didn't finish well. He ran out of fuel, so he had to land in the field, and his plane just turned on its nose. The unhurt pilot climbed out of the cockpit only in a shirt without ID cards and with broken French and asked villagers for a telephone. Although the squadron shot down eight planes, it lost three own pilots, and one returned hurt with a badly holed airplane. After France's fall, the unit was ordered to fly to Oran and Marocco. He is one of the decorated pilots who received the French war cross, Croix de Guerre.

Then it all went very quickly, and on board the ship *Gib-el-Dersa*, he arrived in Liverpool on July 16, 1940. A week later, he joined RAF and was sent to 6.OTU to master the skill of the British plane, the Hurricane. It went very quickly, and František still managed to participate in the Battle of Britain. The following year his 310th Sqn was sent to Aberdeen and re-equipped with Spitfires. A bunch of local photographers and journalists visited the squadron base in Dyce. In Burda's diary, it was commemorated with the words."

"We were visited by 15 Scottish journalists accompanied by photographers. Our bigwigs organized this to introduce Czech airmen to Scottish inhabitants, notably in Aberdeen. Unfortunately, the news leaked that they think the Czechs are a bunch of lazy sods who don't do their job properly and deliberately thwart the war effort of the British public. Obviously, these rumors came from

František Burda - first from the left in POW camp

hostile saboteurs to discredit Czechs. So the publicity event organized to put the matter in order was the correct action. The pressmen arrived at 11:45 were welcomed by the commander and invited for lunch with squadron members. Then we showed them our aerobatic skills and demonstrated dogfights and attacks on bombers. I think they left pretty impressed."

Death was in the back of the mind of the airmen, and sadly it took its victims also during exercise, as is evident from another note in Burda's diary:

"We trained flying in various formations, and I led one section and František Doležal another. We were used to flying in tight units with wings' ends about 15 centimeters from each other. The wind blew from the sea, and the nearby hills produced whirls. About 50 meters from the airfield, the wind threw František and me up, and I spotted how my propeller bit into his rudder. I slowly pulled out the propeller and told him into the radio:

"František, I cut your rudder"

CHAPTER 16: FRANTIŠEK BURDA

Our communication got to the airfield wireless, and all froze there. While Englishmen didn't know what happened, ours did know.

"After landing, we will park on the other side of the airfield so we can examine the outcome of this," I told him.

When we landed and got out of the cockpits, I saw that the wooden blades of my propeller were without the edges while František's rudder was well cut. Thankfully, we escaped unharmed. That could not be said about another incident a few days later.

The glider with infantry landed at our airfield and when it was empty, it was pulled out into the air, and we trained attacks. Approaching it at less than 600 feet (200 meters) was forbidden. It was a clear blue sky, and only one dark cloud appeared south of the airfield. Vladimír Zaoral attacked the glider and pulled the plane straight to the cloud. We all saw him falling in the corkscrew a few seconds later and hitting the ground. The aircraft was smashed, and our friend was killed instantly."

During the war total of 213 Czech airmen were killed during crashes not caused by an enemy. František Burda remembered such bad days in his diary:

"Despite being busy with aerial combat, we still were ordered to do tactical exercises. Sadly, it only increased the number of a long list of killed comrades.

On April 12, 1942, we were ordered to practice interception. Three sections headed to Perranporth, Newquay, and Truro. At 2,000 meters, I requested to dismiss and practice dogfights using photo machine guns. When we returned, two planes were missing. Stanislav Zimprich and Stanislav Halama's Spitfires collided, and none survived, but that wasn't the end of it.

At 11:35, I was about to take off for another practice session, but during the start, the strong wind blew and uplifted my plane. I hit the ground head first or propeller first, to be precise. About 30 minutes later, Jaroslav Šála was about to land from practice, and the same thing happened to him. Again, a strong wind blew, and he crashed, damaging the wing and undercarriage. So the actual outcome of this day was two killed airmen, two write-off planes,

Tom Wilson, František Burda's inmate

and two damaged planes. Obviously, it didn't prevent our superiors from other practices."

On August 20, 1942, Fratišek Burda fulfilled his tour of 200 combat hours and had to leave the squadron to rest.

Saturday, February 27, 1943, was a bad, cold day. The squadron was ordered to attack the port of Brest, where they would accompany bombers from the 8th USAF. Unfortunately, one pilot of section B of the 310th Sqn excused himself from duty, so František Burda opted to replace him:

"I flew as the section leader. When we got closer to the French coast, two Focke Wulfs passed through the left wing of our section. Our formation broke up, so my number two joined the 313sq, which flew with us. About 15 miles from the Brest, I got onto the right corner of our squadron and spotted my number, Václav Řídkošil, in steady flight under me. A few seconds later, I saw him again, crisscrossing under me, but I didn't pay much attention to it. Suddenly I heard the explosion, maybe ammunition in the magazine, and saw the square hole in the wing. My plane fell into the corkscrew, and I couldn't control it though the engine was still working. I opened the hood of the cockpit and heard how other planes of my squadron flew away. I was wondering if I would ever see them again. Something held me inside the cockpit, and I could not get out. I recollected how my squadron mate Emil Foit left his cockpit while trapped there. He simply kicked into the control stick, and the plane threw him out.

I tried the same, and I found myself flying in the air in no time. It took me a while before I found the parachute ripcord and pulled it. Suddenly I was floating in the air in complete silence. We flew about 10,000 feet, and I jumped from the same height. The

sea level appeared under me, but the wind pushed me towards the French shore. Sadly, I also saw how soldiers on the ground were running toward my anticipated landing place. Finally, I landed on the embankment of the small stream. Parachute ropes were frozen. Once I landed, I was encircled by German soldiers. I had to give them the colt I had in my boot, but I still had in my pocket a silk hanky, a printed map, chocolate, money, and other things.

František Burda

I excused myself and told them I needed to answer the call of nature. They took me to the village yard with an outside latrine. I got into it and emptied my pockets, and also threw the emergency kit. Then we waited for the car, which took me to the Brest airfield. Germans brought me something to drink and eat, and soon came a young German officer who spoke good English. We were chewing the fat for a while, and soon the older officer came, could be in his 50s, and he started to talk in quite a good Czech and told me that before the war, he served in the Czech army in Olomouc. He belonged to old Austrian officers, which the Czech government had to take after WWI. Although I could say only my name, number, and rank, he told me.

"We know you are Czech."

Obviously, they figured it out according to my number, 82540. Then he took me next door, where German airmen celebrated victory. From their talk, I understood that two planes went in flames. Much later on, I found out it was my number two, Václav Řídkošil, and another guy from the 313th Sqn I slept on the airfield for one night. The next day, one young Feldwebel and an 18-year-old guy introduced as Josef Kozák from Turnov (a town in Bohemian Paradise) told me they would go to Germany by train and take me to a POW camp.

WASTED LIVES OF UNSUNG HEROES

František Burda

We went to Chartres, which I knew from my stay there, and thought about escape. Still, my guards read my thoughts and warned me with their machine guns, followed me in Paris, where we changed trains and people shouted at me and asked if I was okay. Our next stop was in Frankfurt, and I saw the results of the bombing. It was shattering.

All buildings were knocked down, and the houses or what was left of them were only about two meters high, so I could see quite far. I was taken for interrogation to Dulag Luft. Still, he monotonously answered all my questions with my rank, number, and name. Then, after two weeks, I was sent to Oflag XXIB, where I was told that Josef Bryks and Otakar Černý escaped only a few days ago. My stay here wasn't extended, and after about three weeks, I was sent to Stalag Luft III- Sagan."

In the camp, František returned to his old love- the violin- and became a camp orchestra member. (My good friend Tom Wilson, in 2018 last living member of the vaulting team, which helped with Wooden Horse Escape from the same camp some 64 years ago, told me about František- V.F):

CHAPTER 16: FRANTIŠEK BURDA

"I met František in Stalag Luft III. We both played the violin and were chosen as the first fiddlers into the camp theatre orchestra. The theatre was made from part of block number 64 in the summer of 1943. The orchestra had its place between the audience and the stage. It was not a very big place dug in the ground and exposed to all weather conditions. Due to lack of space, we were lined- up in two levels, us players in the upper one. Sadly, there was also a trumpet player with us, whose instrument was facing my left ear. So when we were all playing together, it was sometimes pretty challenging to observe František's playing though he sat beside me. But even under these conditions, we played from the heart, and with effort, it showed.

František was a great violin player; he was our band leader, and I wanted to emulate his playing skill, so I trained and trained and trained since I didn't want to fail our performance. Together we played music for the theatre play "Thunder Rock," the dumb show "Little Red Riding Hood," and the revue "Bubble and Squeck."

I really liked František, and it was a pleasure to play with him. I can't judge what flier he was since I met him in the camp and I was a navigator in a bomber which was a different ball game. What I know for sure is the fact that he was a very reliable guy, and we could show him any music. He just put his violin under his chin and started to play".

In June of 1944, Gestapo found Czech airmen in prisoners' camps and ordered them to take them to Prague to Petchka palace. Here they were, waiting for interrogation standing face towards the wall. It was here where František was told that one floor above him was kept Bedřich Dvořák, Ivo Tonder, and Desmond Plunkett, who escaped from Stalag Luft III during The Great Escape in March. From Prague, František was taken with other Czechs to Stalag Luft 1 Barth and later to Oflag IV C Colditz, where he was liberated. Three days later, he was already in Britain.

He remarked in his diary:

"I will never ever forget the arrival and welcome in Britain. Firstly, they had to delouse us. Imagine giant hangars with tables decorated with white tablecloths and anything you could think of on

Author with Tom Wilson

Author with Tom Wilson

them. They sat us by the tables, and WAAF came and asked us what we would like to eat and drink. The whole time I have been behind wire- 25 months in total- I have dreamt about a "real English breakfast" of fried bacon with eggs. So, I asked for that, and I ate it. When I finished, they came to me again and asked if there were anything else I would like to have. So, I told them to bring me the same thing again. But I should not have done that.

This was a real treat, while the food I got used to in the past two years was pretty poor. I lost weight in 1944 when Red Cross parcels came irregularly, and we were fed only beetroot soup and boiled potatoes. So when I came to England, I weighed only 54 kilos which isn't too much for my height of 181 centimeters. To cut a long story short, thanks to my first English breakfast after two years, I have spent my first days in freedom on the toilet."

František returned home on August 24, 1945, and half a year later, he married for the first time and became the father of son Jaroslav. But the marriage broke up, and he divorced and married again in 1948 and became the commander of the flying wing in Brno. Soon, he was sacked from the army, moved out from his

service flat, and had to live with his wife and mother-in-law. Also, his wife was sacked from work, and after she found a new one, she was still closely spied on by the Secret Police. Finally, František found a job at a building company in the calculation department, where he worked until his retirement. However, he was still under the watchful eyes of the secret police. Without any reason, he was called for interrogation from time to time. He became the father of other sons, Jiří, in 1952 and Zdeněk two years later.

In the mid 60's, he was partially rehabilitated, and his rank was returned to him. He became a colonel. Since he was 55 when he retired, he has worked as a part-timer, bookkeeper, and translator of technical expert literature. Unfortunately, he died on February 23, 1988, so he didn't live up to full moral rehabilitation.

17 Chapter

JOSEF HORÁK
JAROSLAV STŘIBRNÝ
Two Men from Lidice

Jaroslav Střibrný and Josef Horák

On June 10, 1942, the world was shocked by the bestiality of Nazis who encircled a small village near Prague called Lidice and claimed that they had irresistible proof that inhabitants helped the paratroopers who assassinated Heydrich. They murdered all men, all women were sent to concentration camps, and all minor children were sent to German families for the Reich's upbringing. The village was flattened with explosives, and the name was supposed to be erased from the world maps. But it had the opposite effect. It lifted a wave of solidarity, and some villages worldwide were renamed Lidice. It also increased the determination and the hate of Czech soldiers fighting against Hitler on all fronts.

During that massacre, the men were taken out and shot in front of the wall of Horák Estate. It belonged to the family of Josef Horák,

who was a member of the 311th Sqn and who heard that terrible news from his mate on the airbase.

He was born on June 24, 1915, in Hřebče near Kladno but lived in estate NO.73, which belonged to his father. He was a good student and wanted to attend Flying school, but his parents were against it. So, he started his military career in the infantry, and from here, he was transferred to School for youth pilots in Prostějov. After graduating as a pilot, he was sent to the observation wing in Slovakia. Unfortunately, during one of the observation flights above Polish borders in 1938, he was shot at by Polish soldiers. When he landed, his plane had 28 holes.

After demobilization in 1939, he got a job in Prague, where he met and befriended Josef Stříbrný and also Václav Študent, who later joined him in the 311th Sqn. Two weeks after they left the country, their parents reported them missing at the local police station. The officer advised them to wait another month before coming again since they could still have been in Protectorate and the general search could still find them.

They all went to France via the so-called "Balkan journey" via Slovakia, Hungary, Yugoslavia, Greece, Syria, Turkey, and Lebanon. While Stříbrný went to infantry, both Horák and Študent joined French Air Force. When France collapsed, the former got permission from Czech general Karel Janoušek and boarded Armstrong Whitworth Ensign with 29 other Czech. After a 5.5-hour flight, they landed in Hendon on June 17, 1940, and became one of the first Czech airmen on British soil.

On July 29, 1940, he became one of the first members of the newly established 311th Sqn. The beginnings were much more demanding than in fighter squadrons, since airmen had zero experience with bombing, and the quality of knowledge of the flying staff was very different. Horák was trained as observer, but there was a lack of gunners, so he was quickly sent to a gunnery course. He finished the tour of 200 hours, which meant 35 combat missions above enemy territory. After that, he went to Elementary Flying School to become a pilot. He received his much-wanted wings on January 24, 1942, and returned to the 311th Sqn a month before

Josef Horák wedding photo

that, he married 18 years old Winifred Mary New from Straton near Swindon.

At that time, his squadron was already flying in Coastal Command. He later did another 40 tours above the Bay of Biscay and the Northern Sea. It was on June 11, 1942, when his friend Václav Študent came to him and revealed the news:

"I was getting ready to go to the base when my friend Václav Študent came to me with tears, mumbling two words, "Lidice, Lidice." I couldn't believe that. When someone brought the daily newspapers, and I saw it there, I thought. I was relieved from the air raid and was sent with Josef Stříbrný to the London Ministry of Foreign Affairs, where we met the press. They asked us about our village and wanted to hear our story and the story of our families. We didn't want them to write about us. We knew we would retaliate our way. I felt utter hatred toward the Nazis and almost prayed that none of our bombs missed the target and killed as many Germans as possible. I also felt intense desperation that, apart from short official news, I could not find more detailed information anywhere.

CHAPTER 17: JOSEF HORÁK

It took further three years before I got into my home place, and the outcome surpassed my worst expectations."

Three months after this barbaric act, he was in the pilot seat of his Wellington doing patrol above the Bay of Biscay, when he got the news that his son Václav was born, the first free citizen of Lidice that was flattened. The godfathers were Foreign Minister Jan Masaryk and friend Václav Študent, who was sadly killed not long after that.

President dr. Edvard Beneš was the godfather of his second son Josef born in April 1944. He completed his second tour and completed made 74 missions. After that, he became an officer in Flying Training Command.

His wife learned Czech quite well, and they all returned to Czechoslovakia. Since Lidice didn't exist, they moved to nearby Kročehlavy. From his whole family, Josef met only his sister Anne who returned totally devastated from a concentration camp. During the massacre, she was heavily pregnant, and Germans took her baby, which she never met again. However, his English wife liked life in Czechoslovakia and tried to help Josef build a new home.

He sniffed that it could be dangerous for him to stay in Czechoslovakia. So on April 12, 1948, he escaped to England with his whole family. But even in England, he was under the spying eyes of the Czech Intelligence Service. Then, on January 18, 1949, he got killed during practice flying with De Havilland in lousy visibility. The pilot was killed instantly and was buried at Whitworth Road Cemetery in Swindon.

Josef Stříbrný was born in Lidice on July 28, 1915. After school, he went directly to Infantry School, and when he finished that, he wanted to become a full-time soldier. Hence, he sent the application to the Military academy in Hranice na Moravě. After the occupation, his fate is very similar to Josef Horák. He left home on December 29, 1939:

"I had another shirt in my bag when I left for work. I gave a kiss to mum and dad. They gave me a strange look. That last sight of my parents will stay in my memory forever."

Retaliation message to Nazis - For Lidice

When he got to England, he was sent to a machine gun unit that was part of the Czechoslovak independent brigade and underwent paratroopers exercise. But he was hurt, and the injury put an end to his other career and deployment abroad, so he opted for RAF. He became a navigator and was sent to 1429. Czechoslovak Operation Training Force, where the crews for the Czech 311th Sqn were formed and trained.

The massacre in his home village completely turned his life upside down. He became very revengeful, but the thought pursued his consciousness that he was partly responsible for the tragic end of Lidice. Nazis knew that both he and Josef Horák were fighting abroad and were given a false message that one of them appeared in Lidice during the assassination of Heydrich, which obviously couldn't be true.

His arrival to fully operational duty was delayed by an injury he suffered at the beginning of his active tour. On October 13, 1942, he was part of the crew of Zdeněk Kozelka, who trained in bombing above the sea. During that, their right engine seized. Second pilot Bohuslav Héža who was behind control at that time managed

to avoid hitting the sea level by a whisker and headed to nearby cost. Unfortunately, during a belly landing, he crashed, and the Wellington was written-off. The whole crew was injured, and Josef Stříbrný was the worst case. He had knocked out teeth and broken ribs, which punctured his lungs and caused heavy bleeding. The doctor didn't think he would make it. The convalescence and recuperation took several months before getting back to flying duties. Till the end of the war, he flew in Coastal Command for 1040 hours, from which 605 were operational.

He was rated as one of the best navigators in the squadron. After completing the tour, he was sent to various courses to up his qualification. He took part in the School of Air Sea Rescue, 104Staff Navigation School, and as one of the five Czechs, also Empire Air Navigation School.

He returned home on September 6, 1945, and during a landing maneuver, he flew above Lidice but never dared return there. Until 1948, his destiny was almost identical to that of Josef Horák. He didn't hide his political inclination, so no wonder that after March 1, 1948, he was sacked from the army. His CV stated that he "has a negative attitude to the February 1948 event and doesn't share friendship and love to the Soviet Union. He is hostile towards people's democratic regime".

After release from the army, he was sent to Highest Index Bureau but was under constant observation by the Secret Police. He was arrested and jailed for three months for suspected espionage. They said he verbally insulted the president and got in touch with foreigners. All the accusations were manufactured. Because he was the only living citizen of Lidice, even the brutal communist regime didn't dare humble him as much as other former RAF airmen. Then he found a job in Tesla Hloubetin and met his future wife during one of the business trips. He got married, but through unbearable living conditions, they moved to Písek, where he worked as an office worker, decorator, and stock-keeper. But he suffered from lung emphysema and became invalid. The marriage didn't work out even after the birth of his son and daughter. Nobody

Jaroslav Stříbrný's crew

helped him. Nobody was interested in him. He tried to find solace in alcohol and became increasingly depressed and lonely. He was still pursued by feelings of guilt for the destiny of his home village. He died lonely and forgotten on December 6, 1976, and was cremated. He doesn't even have a grave. His ashes were scattered around Blatná crematorium.

What a sad end for the last living men from Lidice.

Chapter 18

JAROSLAV FRIEDL
From One Hell to Another

Jaroslav Friedl - Liberator, 311 Sqn

Jaroslav Friedl was born on April 16, 1914 in Bezdečí near Trnávka. After graduation, he served his national service and became a mechanic of plane engines. After that, he wanted to become a pilot. So, he started a beginners course in ground mechanics, studying plane technology in a flying lesson. Then, in May 1938, he was sent to Slovakia to patrol the borders. Still, when the Slovakian State declared its separation from the Czech Republic, he took the wing of 30 planes. He flew quickly back to Brno in Moravia.

On August 13, 1939, he left his home country for Poland, where he started to fly Polish planes PWS and protected the airfield in Dublin. When the Germans invaded Poland and bombed his base, he took a plane and flew into Russian territory. Here he was captured by Red Army and sent to a detention camp in Suzdal, where Czech airmen were under the watchful eyes of Russian NKVD. In terrible conditions and often with an empty stomach, he survived 15 months. Finally, on February 21, 1941, the transport was

Jaroslav Friedl standing fourth from right   Jaroslav Friedl sitting second from right

ready for them to take them to Glasgow. The whole cruise took five months, so with his stay in the Russian camp, he spent almost two years on the way to England. He could have spent two years dropping bombs on Germany if it wasn't for the Russians.

In England, he went through Service Flying Training School and Signal School before being allocated to the 311th Sqn. His first operational flight was on December 7, 1942, and Josef Horák from Lidice was in his crew. Firstly, he did patrols above the Bay of Biscay. Still, in the spring of 1943, he was sent to air base Tain in Scotland, and his task was to patrol and protect convoys heading to Murmansk in Russia.

During long and tedious patrols, he spotted and made attacks on Germans ships and submarines:

"On April 18, 1944, we took off from Beaulieu for patrol about Biscay and with landing on Gibraltar. After about nine hours, somewhere at the edge of Spain, we announced the end of patrol and breathed with relief. Then, after casually looking around, I spotted something strange on the sea surface.

"Some strange-looking U-boat is about to emerge under us," I shouted into the radio.

We had never seen such a strange shape and couldn't identify it even after another circle. We were off duty now and hesitated with the attack since we weren't sure if it was a German or Allied submarine.

We made another circle, dropped marking colors, and fired smoking rockets for another Liberator sent here after our call.

CHAPTER 18: JAROSLAV FRIEDL

Author with Mrs. Friedl 1994

When we landed in Gibraltar, the Intelligence Service Officer confirmed it was a particular type of German mine-laying submarine. Two destroyers followed it to Brest port, where it was sunk.

A day later, we had another quite exciting patrol. After refilling and a few uneventful hours, we spotted small subjects under us. After reporting that to our commanding officer, we checked it closely and eventually attacked it. After two circles, we were sure they were four German ships in small convoys. To baffle them, we fired rockets into the trees," which made a nice bonfire. Then I turned the Liberator into a gliding flight with an increased acceleration of about 270 mph to give their guns the smallest possible target and little chance to hit us. It paid off since we got to the first ship and dropped deep charges close to its port. After that, we spotted an enormous wave caused by a mine explosion, and the boat turned over. We didn't return, so we have no idea about the fate of the whole convoy, but we guessed some other patrolling Liberators above Biscay finished it off."

During D-Day on June 6, 1944, the first plane from the Czech squadron, which took off at 5:30 for support of the invasion ships, was the Liberator L 717 commanded by Jaroslav Friedl. He was

Author by the grave of J.F. Mestecko Trnavka

accompanied by two Czech Liberators, E 745, commanded by J. Hala and Q 741, commanded by L. Moudrý. They had a chance to see the breathtaking invasion scene, which spread from one horizon to another.

With operations connected with D-Day, he came close to 50 combat missions. He added another 20 during patrols from the North Sea to Polar Circle. To confirm that these patrols were no mean feat, we can add the comment of the 311th Sqn tail gunner. "Trips above Germany were hell, but flights above the endless water for ten hours were three times worse."

At the war's end and in the first months of peace, Jaroslav Frield flew with transport command and brought home many VIPs, including Foreign Minister Jan Masaryk. In 1946, he demobbed and went to Czech Airlines, where he flew on home and international lines. He also married Zdena Pálová, who was to share the tough life ahead of them.

Soon after February 1948, they both started to be monitored by Secret Police and accused of espionage and keeping in touch with ex-RAF mates. In 1950, he was thrown out of Czech Airlines,

although he was their most experienced and top pilot. He was forced to move out of Prague, so his second home became a town called Písek, where he found a job as a mechanic of medical instruments.

But even here, they were both watched by Secret Police and then arrested on the same day on April 15, 1954, on the pavement and in handcuffs, transported to Prague jail Ruzyně. The accusation was, of course, false and constructed by the Communist Party. However, Jaroslav still got 13 years of heavy manual work, and Zdena got 16 years since she dared protest against false charges during the trial. Hence, she received three years on top. Jaroslav had to work in a uranium mine in Jáchymov and then was sent to famous jail for felons in Valdice. They spent six years in prison each before amnesty in 1960 released them. They returned to Písek and tried to find jobs, but being labeled as "political ex-cons," they were rejected everywhere.

They went to pension in 1969 without even partial rehabilitation. A brutal communist gauntlet known as "normalization" came in 1970, meaning they were both haunted and watched by communists again. They retired and lived very modestly and lonesome at the edge of Písek with minimal resources. It took another twenty years before the regime was beaten, and they could breathe freely. So, after 41 years, they were both morally rehabilitated for the treatment they had received for something they hadn't done.

Staying in jail and working in uranium mines took its toll, and Jaroslav Friend died on October 9, 1992.

19 Chapter

VLADIMÍR SLANSKÝ
Flight Sergeant "The Broom"

Vladimír Slanský After the war

Vladimír Slánský was born on July 30, 1913 in Mladá Boleslav. He studied at s State engineering school before serving his national service stint. Being technically minded, he became a radio operator. In June 1939, he escaped to Poland and asked to be accepted into Polish Air Force but was refused. Instead, he had to sign the five years contract with Foreign Legion. On July 31, 1939, the ship *Castelholm* anchored in Calais, and the group of would-be-airmen went by train to Paris.

It took almost two weeks before all medical checkups and paperwork were done, and they could enter Saint Jean fortress - the home of the Foreign Legion. Finally, after four days, the whole unit was transferred to Sid-bel-Abes, where they all underwent rugged boot camp.

When the war was declared, Vladimír and others were sent to French Air Force and went to radio navigation and gunnery school. Since he already had some experience from home, he was soon chosen as the instructor of young air navigators. But before

all could get into full swing, France gave up, and all went into chaos. Vladimír's commander told them he had no further orders, so now it is "everybody for himself." So Czechs persuaded the captain of the Dutch cargo ship *Ary Scheffer* to take them all on board and dispatch them to England. The Falmouth welcomed them on June 22, 1940.

As a very experienced navigator, he was much wanted in the newly established 311th Sqn and got onto the crew of Václav Korda, with who he also made the first combat mission of this squadron on Bruxelles. It was on September 10, 1940, and meant a big day for all concerned since it was the first chance to pay back the Germans, stand them face to face, and fulfill the wish that everybody escaped his country. Three crews were chosen, and Slánský and his Wellington took off as the last.

As he later recollected:

"Everybody was standing alongside the runway; everybody wanted to be there on the very evening; it was the big day for all of us. Nobody went to bed; everybody wanted to be up and wait for our return. The buzz was felt, and everybody wanted to chip in with some advice. They all told us don't fear, probably because nobody knew what to expect. Of course, we all were scarred, but the stress left once the engines were switched on, and we all got busy with our respective tasks. Immediately after taking off, we got into heavy clouds and flew above them at 1,000 feet. According to the order, we descent onto 9,000 but were still in shadows. So we went lower and lower and still didn't see anything. Then we were far too low not to breach order and bomb an invisible target, the pilot pulled up, and at 21:32, he turned the plane back home - with a full bomb load on board. We landed at 23:25, and everybody came to welcome us, but we didn't feel like that."

We would wish there was nobody on the runway. We couldn't hide our disappointment, and all was clear when the armament king opened the bombs bay.

"Fucking hell, you brought the shitload back home."

Everybody jumped at us and flooded us with various questions. "What happened?"

First Czech radio operators V.S. 1st from left

"Malfunction or you got lost?"

We all felt humiliated, and none of us had the will to answer. Captain had a problem explaining everything to the commander-in-chief. Finally, he had to explain why the order was breached.

When we came to the dining room, nobody had spoken to us, all were distant, and the silence was nerve-wracking.

It all ended up with air gunner and former boxer Vilda Jakš who stood up and told loudly for everybody could hear.

"They didn't drop today; they will drop tomorrow, the day after, and the day after, no big deal."

The next day it was announced that our crew did precisely what was ordered and deserved acknowledgment for that."

Another air raid above Brussels train station followed a few days later. It was again lousy weather and a cloudy sky. When his Wellington got above the target, it was already covered in smoke, and visibility wasn't good. Not to repeat the same result again, they crossed it four times to be perfectly sure. Circle twenty minutes above a heavily defended target is nothing one would really fancy, but their pride was at stake. Finally, they dropped the bombs and hit the marked spot on.

CHAPTER 19: VLADIMÍR SLANSKÝ

Wedding photo

On the return journey, they returned to heavy clouds and flew only 100 meters above the ground. When they landed, they were rightly happy to finally fulfill their task. When the crew got to their beds, lonesome German bombers dropped bombs above the airfield but poorly aimed and didn't do any real damage.

Vladimír Slánský was not only 100% professional but also a lover of spotless cleanness. He bought a broom for his own money and always swept and cleaned the whole fuselage of the Wellington. It had to be spot on. Like in any other office, he put a small curtain into the window next to his radio operator's chair. No mechanic dared touch it with his greasy hand. As time passed, he became known in the whole squadron as Flight Sergeant Broom.

On June 22, 1941, they bombed the heavily defended target of Bremen. During that air raid, one bomber was shot down, one had to land in Holland, and the crew was captured. Slánský crew had Lady Luck on their side:

"The flight at Bremen took 5 hours 4 minutes. The weather was good, and above the target, we got into the worst AA fire I have ever experienced. We had to fly through many such barrages to get exactly above the target. The air was full of grenades. The

explosions were literary everywhere. The target under us was in a fire, and smoke covered it. To be accurate, I had to tell the pilot to do one more circle just above the center of that hell. We were above the target for about twenty minutes, and I wondered why they didn't shoot us. Once we dropped the bombs without hesitation or trouble, we flew back to base. After landing, I opened the exit door, and something hit the ground with a bang. It was big shrapnel with a razor edge. We realized it entered the fuselage on the left side and missed the bomb aimer by about twenty centimeters. When he lay on the floor aiming at the target, it missed me by a whisker and hit the metal construction of the plane on the other side before dropping down on the floor."

Vladimír Slánský flew 45 operations, in a total of 212.40 hours. After that, he was sent to the 138th Sqn. For special duties, he made three trips here, one to France, one to Czechoslovakia, where he had to return for lousy weather, and one to Norway.

He always wanted to be a pilot, so he applied for pilot school. First, he was sent to Canada, where he went through Elementary Flying Training School near Calgary. Then, in Charlottetown, he also went to school for navigators. Next, he was sent to the Bahamas to O.T.U., where he learned to fly Mitchell and Liberators. Then, as a very experienced airman, he was sent back to England and returned to his 311sq, which was now flying as a part of Coastal Command on anti-submarine patrols.

He stayed and survived there until the end of the war. His total tally at Coastal Command was 380 hours during 36 missions which wrapped up his war score to 84 tasks. After the end of the war, he became a pilot for the Transport command and flew from London to Prague and back, bringing personnel, material, and archives. When he demobbed, he was looking for a job. His experience and education made him a much-wanted person for Czech Air Force. He flew on international lines for four years.

In September 1950, the airlines immediately sacked him immediately after returning from Sweden. His "sin" was the fact that he flew in RAF during the war. He found a job in Mladá Boleslav as a grinder. In 1953, he was arrested on the way to work and

CHAPTER 19: VLADIMÍR SLANSKÝ

One year before death

sent to jail. The charges were manufactured, and he received six years in prison and confiscated all his property and possession. He was kept behind bars in Prague and Jáchymov in uranium mines and Leopoldov-notorious prison for felons. Such was the treatment of the man who received four times Czech War Cross in 1939, twice the Czech Medal for valor, and numerous Czech and British awards. He outlived the much-hated communist regime and was morally rehabilitated in 1991. The freedom to travel enabled him to not only visit England and France many times and meet his war mates but also get to Florida and visit his captain Václav Korda.

Vladimir Slansky died of bladder cancer on August 24, 1993, at the age of 80.

(We befriended, and I can only say that he was an amiable and modest man and stayed that way till his end - V.F.)

20 Chapter

RICHARD HUSMANN (FILIP JÁNSKÝ)
Heavenly Rider

Richard Husmann

The following life story could serve as an excellent script for a film about an adventurous man named Richard Husmann, known by his pen name Filip Jánský. Born on September 4, 1922, in Prague, Richard was a scout and was taught fencing, horseback riding, and English from a young age. Major Stanislav Mašek, a lodger in Richard's mother's flat and a prominent figure in the Czech resistance movement against Hitler, chose him for intelligence work when the Germans invaded Czechoslovakia.

Richard was given the task of visiting Germany to observe Braunschweig, the center of the second aviation army, and to visit North-West ports such as Emden, Wilhelmshaven, Bremen, and Lübeck, with a focus on submarines. He was also to visit the Siegfried line as a tourist to see if German soldiers occupied it. At 17, Richard left the Czech Republic in June 1939 with little money and only wore shorts and a shirt.

The young man, who had never been away from Prague, faced many challenges as he traveled alone to fulfill his espionage tasks.

Yet, despite his scouting attire, he could stay overnight in homes for Hitlerjungen for just one mark, shower, and have breakfast in the morning. He even managed to get a job as a waiter assistant in Harzburg and earn some cash.

After informing his Prague contact, Richard moved to Belgium and France by train. He visited the French consulate and was sent to a Czech summer camp, from which he applied to join the Foreign Legion. Unfortunately, he added two years to his age and ended up in the St. Jean fortress in Marseilles. Although he admitted that boot camp was brutal, he probably would have deserted if it wasn't for the outbreak of war. However, due to the war, Czechs were released from the Foreign Legion, and Richard could join the Czech army in Agde.

Overall, Richard's "I don't care" attitude and undeniable bravery, despite his scars, made him a one-of-a-kind character among Czechs in the RAF.

At 17, he was France's youngest Czech soldier and later the RAF and 311th Sqn member. In Agde, he was in the same unit as Jan Kubiš, who went down in history three years later as one of the brave paratroopers who assassinated Heydrich. Josef Gabčík, the second brave paratrooper, was the leader of a nearby machine gun unit. He also met another paratrooper there, Karel Čurda, who betrayed all his mates and sent many Prague helpers to the hands of the Gestapo. Another notable man he met in Agde was Augustin Přeučil, an airman who joined the RAF, made debts, got into trouble, and flew away with a Hurricane to Belgium, where he gave it to the Germans. He became a traitor, worked for the Gestapo, and identified the shot-down Czech airmen. After the war, he was found guilty and executed. Richard stayed in Agde for eight months before France gave up. He was on the run again to Port Vendres, where he boarded the ship *Apapa*, which headed towards Gibraltar and then to the English shore.

The Czech soldiers didn't know then that the Germans, who had occupied Paris, raided the Czech consulate and confiscated the complete undestroyed files of all Czech soldiers from the exercise base in Agde, which included the airmen. Other files had the

Richard Husmann shot from film Heavenly riders made according to his book

names of about 3,000 relatives of Czech emigrants. They all went to the Gestapo in Prague. When the Czech airman-traitor Augustin Přeučil flew away to Holland with his Hurricane, he had files of all Czechs in the RAF, including their ID numbers and ranks. When Czech airmen got captured, the Germans immediately knew who they were, making them very vulnerable. So, they needed to be in the RAF and swear an oath to the English King, as it protected them from the Geneva Convention.

Richard was accepted into the RAF and underwent boot camp. The Czech 311th Sqn got English instructors. One of them was the legendary Charles Pickard, who was killed during a brave attack on a prison in Amiens in 1944. The second was New Zealander Tom Kirby-Green, who later became a POW, escaped during The Great Escape and was murdered by the Gestapo in Ostrava, Czechoslovakia.

Richard soon adapted to life on and off the base and didn't hide his adventurous character.

He enjoyed poaching:

"I was pleased with the move from Honington to East Wretham. We used to get to East Wretham by bus before the Manor farm was

CHAPTER 20: RICHARD HUSMANN

hired since the living quarters were appalling. Not far away was a pheasantry, wild hares were plentiful, and the local ponds were rich with carp, spikes, or eels. We found good resources to enrich our monotonous base menu, and we believed that all our spawning sins would be excused on a doomsday. My culinary skills became well-known soon, and I was happy when I could fill the stomachs of my friends. But I wouldn't wish anyone to undergo those moments when I prepared a full pot of tasty food for my mates, and they didn't return from the mission that night." Since the pub in Bury St. Edmunds was too far for our liking, we all chipped in and bought battered cars to take us directly to the town. These cars often looked like buses in India, always carrying extra bodies. The main challenge was the lack of petrol. Still, we soon found a solution by using petrol for bombers, which was strictly forbidden and controlled. The petrol had a greenish color and was readily detectable in the engines.

To avoid trouble with the "bobbies," our mechanics built a second tank and installed a switch on the dashboard. During our rides, one person was assigned to watch duty. If a police control was in sight, he would give a warning, and the driver would switch from the "plane petrol tank" to the proper one, so nothing could be spotted if the bonnet was opened. Our mechanics, known as "The Golden Czech hands," demonstrated their skills even on British soil. I shared the car ownership with two others. One of them, Zdeněk Babíček, was my roommate but ended up in a Wellington wreckage near Wilhelmshaven. The second co-owner drowned in a crash at sea, leaving me lonely in the car on my way to the pub. On my way back, I crashed into a masked tank, and the car was a total loss, so I got out and walked home without looking back at it.

One day, on December 16, 1940, I was sitting in the airbase club room, reading the papers, thinking about my killed mates, and waiting for an English friend. Then, around 18:00 a loud detonation was heard. Everyone pricked up their ears, but when nothing else happened, they all returned to their beers.

Then the door opened, and several sergeants dressed in flying overalls walked in. Somebody asked what had happened, and one, a gunner, according to the sew-on patch on his overall, said:

"One Wellington crashed during takeoff and is in flames. It was that fucking Czech bastard. The takeoff was aborted."

I felt insulted when someone commented on the death of my mates in such a manner, so I dropped the papers, got up, switched off the radio, and shouted:

"If there is any fucking bastard, it's you."

I switched the radio back on, returned to my chair, and waited. The sergeant wanted to attack me but was held back by his mates. So, he asked me to go outside and have a fight. While the English were used to boxing, we Czechs were used to slapping faces. But my opponent forgot about his triple underwear and heavy woolen boots, making the fight unfair. I danced around him and threw my right hand, directly hitting his nose. He started to bleed and didn't want to continue the battle, so we all returned to the club room. I walked in last and told everyone I was Czech and could make as many ops as they wanted whenever I wanted. That was it for the night, but it had unexpected consequences. One afternoon a car came to our sergeant mess, and "my gunner" walked in with his friends. They asked me to go out, and when they assured themselves that nobody listens, they offered me a trip in their Whitley above the Reich this evening.

I agreed under the condition that the gunner apologized to me for the last insult, so I picked up my helmet and gloves and prepared for my first operational flight. I told my roommates Skalický and Kadlec, who found their death in the sea before the middle of the war that I had a date and went away. That gunner offered me more such trips, but one evening he didn't show up to pick me up, since he was shot down."

Richard didn't like authorities and was in a personal war with squadron leader Josef Schejbal. It started all during the parade on March 7, 1941:

"The whole squadron stood in attention, and to the absolute silence, I shouted loudly Aleeeeeeeeert!!!!!!

All ran away since they knew from experience what it was like when enemy planes bombed the airfield.

Commander called me into his office and gave me a hard time for dismissing the parade. I admitted innocently that I heard some

CHAPTER 20: RICHARD HUSMANN

France 1940 - Antonín Kriz, unknown, Richard Husmann

droning. It is publicly known there that he doesn't hear properly and that pre-warning eluded him.

"Can you imagine the results of exploding bomb in the shut hangar? We would lose half of the squadron," I said without a blink of an eye.

The heated discussion followed till he concluded with a remark: "We don't need either cowards or D'Artagnan."

Since that time, I have ignored him. Although I saluted him correctly when he talked to me, I didn't answer him.

After a while, Josef Schejbal gave up that battle, but he had the last laugh. When I was suggested for DFM, he didn't give his recommendation. I was upset, but only for a while since medals and awards didn't protect anyone from death."

When Richard was in the navigator's course, he was called by flight commander Captain Josef Ocelka who friendly asked him to go to Leamington Spa, where the base of Czech infantry was. He was to spread the word about the good life in the bomber squadron and try and draft as many new gunners as possible. But, unfortunately, he was discovered and had to return. Still, about fifty new recruits soon came to the 311th Sqn so his action was fruitful.

When the operational flights became regular and the bombing of Germany got into full swing, Richard realized what it really meant

to fly above heavily defended cities and face death daily…and he started to fear. He openly admitted that, unlike others who tried to sweep it under the carpet, sometimes with glass or bottle of spirit.

"After the raids above most feared targets Cologne, Düsseldorf, Essen, Ruhr valley where I was looking into the opened mouth of hell, I now realized what it is all about. I started to suffer terrible fear, and I knew that if I didn't beat it myself, no one else would help or save me. I, of course, could have refused to fly, and they would strip me off my rank for lack of moral fiber and sent me to the infantry. I would have been ninth already. I didn't toy with that idea, but I knew what frightened me the most. Fire. When we went to bomb Hamburg, we took 640 gallons of fuel. If the tanks placed in the wings got a direct hit, it was followed by an explosion, and the whole wing was chopped off. The plane's body started to fall and roll, and the flames engulfed it. The crew had almost zero chance to bale out.

At the beginning of the war and in older types of Wellington, the fire inside the petrol tank couldn't have been extinguished. The engine had its own fire extinguisher controlled by the pilot's cockpit. It was foam that killed the flames, but after that, the engine was usually in such bad shape that it couldn't ignite and work again. Once the fire got inside the fuselage, it was the end since there was so much food for it-ammunition, flares, and machine gun belts. Sometimes, the turret door got jammed, and the gunner was burnt to death.

The inferno and flames on board caused utter horror, which an ordinary human being couldn't outlive, the animal self-preservation instinct prevailed. We have heard that some of the crews shot down above Hamburg, who managed to bale out, landed on the ground, were caught by furious inhabitants, rolled into their own parachutes, and thrown into the first fire caused by the bombing. Local Mayor Kaufmann told Hamburg citizens to fight one terror with another.

But he has forgotten who started the war…

I slept poorly, ate badly, and used to go for lonely walks with the gun in my pocket.

CHAPTER 20: RICHARD HUSMANN

We got Czech Valor Medal for three operational missions and were nominated for Czech War Cross for twelve tasks.

Once there was a parade, and we stood in attention. Commander Josef Ocelka read the names of decorated men, and I was among those to get the Valor Medal. I got the idea that I would fight the fear on the ground first. So after hearing my name, I said loudly:

"Commander, I refuse to accept that medal."

Commander stopped, lifted his head from the list, and asked me:

"Can you tell me the reason for that, sergeant?"

"Reason, Sir? I am not brave. I have fear," I replied.

While the commander was furious and my action didn't meet with the overall understanding, I now felt calm since I knew that I had won my first battle with fear.

Not long after, Richard had another strange experience that marked him for his future life. Most people at least once experienced something they couldn't explain realistically and labeled it as a sixth sense. The same happened to Richard in the refreshing gunner's course in Newmarket.

"I was sent to the course with my friend Vladimír Zapletal. We were in the second week and were about to go for live firing. We walked towards the base, and something strange happened to me. I stopped and told Vladimír that I quit the course and won't fly and leave the base immediately. He agreed although he was baffled by my swift decision."

"I told the commander my views on the course, and I packed up and returned to my unit. I added something that surprised me."

"Sir, the plane we will fly with and practice live firing will crash in the afternoon."

He didn't believe his ears and asked me to repeat what I said. Then he started shouting at me and threatened to contact my commander. I just shrugged my shoulders, saluted him, and left.

When I got to our base, I was called to Commander Josef Ocelka, who looked at me with a face that looked like he was eating a lemon. He asked me about the course, and our conversation was interrupted by the phone. He picked up the receiver, listened to it, then put the phone down and said nothing.

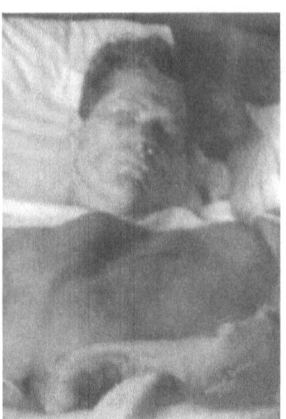

Richard Husmann injured 16.4.1945

"You can go sergeant...no, hang on...I am warning you, if you tell anybody in my squadron that his plane will crash, you will see off the war in jail."

"I saluted and left, knowing what he was told on the phone. The squadron grapevine worked, and soon I was asked by my mates how did I know that. But I couldn't give them a good answer. I have never had premonition feelings, but since that time, I always followed those instincts."

During the crash landing from an air raid above Cologne, Richard got injured and suffered a bad concussion. After recovery, he got involved in another adventurous occasion.

"Once, someone informed police that after the flight cancellation, the Czech bomber squadron entered their cars and went to pubs. All our vehicles had those special dual tanks. Police cars blocked the road and took the samples from the tanks. They had a bag of victims, including seven plane captains, a wing commander, and the squadron commander. It was a massive problem reported to the Czech Inspectorate in London. Its boss Karel Janoušek wanted to find a peaceful conclusion to that case, but Brits didn't want to hear.

Its results would mean a big problem for the operational duties of the whole squadron, and President Dr. Edvard Beneš would have to be informed. It would put us in a bad light. Since the samples were still kept in the base commander's office and waiting for the special investigation committee, I got there with gloves on and stole all models. Scotland Yard was called, but nothing was found. Reports with car owners weren't signed yet, so further investigation was scrapped, and our squadron was kept operational. I must admit that I received a blessing from our commander Josef Ocelka.

During the crash landing from an air raid above Cologne, Richard got injured and suffered a bad concussion. After recovery, he got involved in another adventurous occasion.

CHAPTER 20: RICHARD HUSMANN

"Since that return to operational flying, I have been scared. I felt dizzy, couldn't swim, didn't mix with people, and was nervous. I thought it was a sign of a coming end. When climbing into the plane for the raid above Mannheim, I accidentally pulled the parachute opener, which opened as expected. Unfortunately, we were about to take off. We were on the other side of the airfield, so it was impossible to grab another parachute. I could have refused to fly, but instead, I chose to take off the parachute, drop it to the ground, and fly without it. The plane captain, Metoděj Šebela, looked at me in confusion, but then he shrugged his shoulders, and we took off.

Before we reached our target, I noticed a shadow behind us and reported it to the captain. He changed our direction, and so did our visitor. Suddenly, the plane climbed to 4,000 meters and increased speed to prepare for an attack. I turned my turret and fired a long burst.

"The enemy's left engine is on fire," I shouted over the intercom and felt sweat on my back. The front gunner confirmed that he saw the enemy's plane in flames and fired a burst as the enemy flew beneath our Wellington I didn't even realize we dropped the bombs, and I had time to reflect on the fight.

I realized that I wasn't scared at the moment of the attack. I wasn't afraid to fly without a parachute, and I wasn't scared to face death. I understood what life and death are all about. We will all die someday, but we must fight against it and not be afraid.

I was not credited for the victory during the intelligence briefing after the flight, despite the front gunner confirming it. This was because we didn't have camera machine guns in our turrets, which made me so angry that I left the room without saluting. When I returned to my quarters, I thought about that night and realized I felt no fear. I decided that from that day on, I would always fly without a parachute.

From that day forward, I changed and no longer saw my superiors as authorities. Of course, I talked back to them, which cost me my sergeant's stripes, but I eventually got them back."

Czech 311th Sqn was transferred to Coastal Command and flew patrols above the sea. Richard continued with his "hobby"

and flew illegally with foreign crews on missions. When any gunner didn't feel like flying, he would ask Richard to swap places just before the flight so nobody would find out. He flew over 300 hours above the sea.

Sometimes, he chose the captain and his plane and slept either near the plane or in the fuselage with only a flask of water and a gun in his pocket. During flights, he would sometimes appear in the pilot's cockpit to say "hello" or announce over the intercom that he was at the side machine gun on patrol. Although it was surprising for the plane captain, everyone was happy to have him on board as he was able and willing to replace anyone in any position, whether it be a gunner's place or the navigator's chair.

He flew without a parachute or Mae West to overcome his fear. He was like a happy, lucky charm for everyone and was said to know which plane would return and which wouldn't. Once, Richard was supposed to fly with the crew of Ota Žanta, along with his friend and gunner Ladislav Kadlec, who had lost a leg during an air raid in 1941 but still continued to fly with an artificial limb. However, at the last minute, Richard declined the flight, and the plane never returned from the Bay of Biscay, causing Ladislav to end up in the sea. After that, Richard was known to choose his own flights and not fly with just anyone who asked him to.

There were often problems when the plane had to land at different airfields for various reasons. The operational officers reported fewer crew members than the official records. This led Squadron Commander Josef Šejbl to release a particular order, requiring every captain to check the plane before the flight with the whole crew standing by. Richard sometimes slept above the bomb bay, covered with canvas.

Richard "guest flew" with the squadron during the eight months, and no plane crashed during takeoff. However, after he left the unit in June 1944, seven planes crashed during takeoff in the following ten months, all skippered by experienced pilots. Many probable causes of sabotage were never explained. As a result, Richard often held guard and slept under the fuselage with his gun. He even spotted suspicious acting Czech and English mechanics

Programme for film Heavenly riders made according to his book

a few times. He discovered an airfield dog with a bullet hole in its head on a nearby beach.

The cockpit was a very tiny and uncomfortable place. The seat was from canvas or plastic, with no armor plating to protect the gunner. So, he was basically exhibited to enemy bullets. Apart from this, there was no central heating, so it was freezing inside.

They got into combat missions, slowly crossed the borders with Czechoslovakia, and fended off Germans in Northern Moravia near Ostrava. Then, only three weeks into the end of the war, on April 16, 1945, Richard had another meeting with death. His plane was attacked by FW 190 from the back, and his cockpit was holed by accurate fire. He was severely injured and heavily bleeding, but he still managed to shoot it down in flames. Richard was taken to the military hospital. Since he was in shock and the doctor didn't know his blood group, he was operated on without anesthetic.

For three days, he balanced between life and death. He suffered multiple splinter injuries on his face, head, and limbs. One of the FW-190 cannon shells hit the stand of Richard's machine gun at the very moment he was aiming. It disintegrated into pieces, and he was happy to be alive. One centimeter higher and he would have been decapitated.

The war ended, and he beat death in 54 operational missions. He returned at the age of only 23 with no education. He had tempting offers to stay in the army but declined them all. Finally, he had enough of the uniform, which he donned for 2130 days.

The first year after the war, he worked for UNRRA. Then he worked at Prague airport in the control tower. He wasn't very well off and rarely spoke about his war experience. He wanted to study at university, and despite speaking English and Russian, he was expelled in 1951. He barely found a job as an assistant worker in Prague. After a year, he started working as a translator and then

in the literature business. Due to war injuries, he was entitled to an invalid pension of 252 crowns per month (8 pounds in today's money). Later, he was reduced to 87 monthly crowns (3 pounds in today's money).

He became like a snail, living in his shell and avoiding people. He burnt out all his war mementos, and there are only two known photos from RAF time these days. In his mid-60s, he wrote his most known novel, "Heavenly Riders", which sold 147,000 pieces. In 1968 the film of the same name, "Nebeští jezdci", was made according to his book by director Jindřich Polák. Some ex-RAF personnel worked as supervisors. Richard Husmann, aka Filip Jánský, was very rarely at the shot.

The director had very little money and needed an English cast. When he explained to the actors that he couldn't pay them much, they said, "Don't worry, we owe you this." He was loaned reels of war documents in an English archive which he put into his car boot. The whole archive caught fire shortly after, and all documents were damaged- all but those in the boot. He shot in England only for two days. The rest was done in Czechoslovakia. The Wellington was made from Dakota, and feature parts were so well edited and mixed with documents that on the premiere, former RAF airmen were guessing where it was shot and swore they served there during the war. The last take was done on August 21, 1968, the day Russians invaded Czechoslovakia- how symbolic.

The premiere was banned from Prague cinemas, so it was organized in Kolín, about 40 kilometers from the capitol, in December 1968. Many airmen were invited but were already closely observed and followed by Secret Police. It only had limited screening since the following year. It was banned from the cinemas and locked in a safe for 21 years.

Authors note

My father lent some war items to a film studio and got them back after a decade. They are seen in the movie. I personally think this is the best war film ever made. When I sent copies to my English friends from ex-RAF personnel, they told me, "that is exactly as it was during the war."

Richard Husmann gunners course, Dumfries 1940 - Husmann first from right

Richard Husmann died two weeks before his 65th birthday on August 20, 1987, he didn't live long enough to get moral rehabilitation. In his modesty, he once wrote:

"Why did I leave my country at the age of 17? I thought that was right; our fight didn't stop with occupation. Of course, I was very young, but 17 years old man takes things seriously, often more seriously than those older ones."

21 Chapter

KAREL ŠEDA

From Hero to Zero in One Decade

Karel Šeda

Karel Šeda was born on November 4, 1908 in Ujezd near Chocen in Eastern Bohemia. He became an electrician apprentice but loved flying and became a pilot in 1929. His flying abilities were spotted by experts of the flying company VTLÚ Letňany, and he became their test pilot. He couldn't complain since his monthly wage was 1500 crowns (50 pounds in today's money) which was a lot of money in the pre-war time. He became a pilot for Czech Air Lines in 1935. Firstly, he flew on domestic lines and then was promoted to international ones. When war was declared, he was offered a chance to join Luftwaffe but obviously declined and escaped to Poland and France. He managed to fight as a member of the French Groupe de Chasse II/2 and flew in the French sky for 21 combat hours. Then, when France put down arms, he escaped on the ship *Gib-el-Dersa* to Gibraltar and England. Three weeks after putting his foot on English soil, he joined RAF as a sergeant and was welcome to the newly established 310th Sqn.

During the Battle of Britain, he had one damaged Messerschmidt Bf 110. (At this time, he sometimes visited my grandparents in Cambridge, and they became friends- V.F.)

He became a test pilot in various Maintenance Units. After two years, he applied for operational flying and was accepted to the newly established night Czech- British 68th Sqn. He was classed as an "above-average pilot with tons of experience."

Firstly, the squadron was equipped with Beaufighters and later replaced by excellent Mosquitos. In 1944 there weren't as many German aircraft in the British skies as there used to be in 1940, so he could add only one confirmed German plane to his tally. So, he became the only Czechoslovakian airman who shot down Ju-88.

He explained the difference between daily and night fighter pilots:

"Life of the daily pilot depends on the flexibility of his neck. He must look around and also use the mirror. The sunshine at 8,000 meters is blinding, and there is still a moment of surprise. Germans attack from the sun as lightning. We constantly attacked the enemy with visual contact, sometimes head-to-head. When our Hurricanes had to attack bombers, quicker Spitfires were to protect us from enemy fighters. When we all got together, it was a big mêlée and planes were left, right and center. It was incredible. We had to look in front of us to aim at the enemy and be careful not to hit our mates. Also, we had to look around not to become a target for the enemy, so it was big adrenaline rush.

When night fighting, we have radar to look around. Sometimes, we can get carefully behind the enemy without them spotting us. Usually, we shot them down with the first burst. The second is usually coup de grace, just to be sure. At night there is still a significant danger of mid-air collisions even with a friendly plane."

Squadron chronicle remarked on the day when the squadron got the news about being equipped with Mosquitos:

"The bar was opened in sergeant's mess in the morning. All were drinking like a fish. The dinner was attended by all members of the 68th Sqn. At about 9 p.m. (2100) the squadron commander and about ten others decided to go to a local pub. Transport of 11

Karel Šeda, president E. Beneš, Rechka, unknown

people in the four-seater car was with problems. Captain Hickin and Pilot Officer Cupák had to make do with the boot. The party was very merry by Adam and Eve, and the return journey was much more comfortable since only nine people wanted to get back."

After finishing his operation tour, Šeda requested a transfer to Transport Command. With 4,600 flying hours under his belt, he didn't have to go through training and could start with Metropolitan Communications Squadron immediately. He returned home on August 17, 1945, and was a pilot in Czechoslovakian Air Lines for four years.

In March 1950, he was sacked from the airlines, and it was almost impossible for him to find a job as a former RAF pilot. Former "Westerners" met up occasionally, and they pondered what to do. One of them proposed that he had a friend named Karel Úlehla, who would take them with their families abroad. But Úlehla was an agent of the Secret Police, and he revealed the whole affair. Karel Šeda was arrested on December 23, 1949, and within five weeks, there was a communist trial. The manufactured lawsuit accused him of treason and conspiracy against the socialist republic. He was sentenced to 14 years in jail for hard labor. Ironically,

Leopold Formánek (author's father), Miroslav Jiroudek 68 Sqn, Karel Šeda 68 Sqn, 1990 meeting after 50 years

he got the highest charge but did the least from the whole group. From Prague prison, he was sent to jail in Plzeň–Bory and from here to Horní Slavkov in Moravia, to the uranium mine.

He recalled those days:

"I will never forget that corridor made from barbed wire divided camp from the mine. We were forced to run, tied if fives, and on the side were standing guards with machine guns. We had to work at the devil's speed. In 1955 I was sent to Jáchymov, a uranium mine. The working conditions were like those when pyramids were built. It was a concentration camp with everything apart from cremation chimneys. Many people didn't know about the existence of these uranium mines, but in Jáchymov, there were twelve of them, and in them, 25,000 enslaved inmates. "

"I was lucky that later on, I could perform my trade, electrician. Soviet experts, in fact, ran those mines, and all uranium was sent to Russia as a friendly help to "create peace." I was released on probation after eight years. We asked for a retrial. Finally, we were all fully rehabilitated on November 4, 1966, as it was

Karel Šeda in Duxford L-R Fürst, Rechka, Seda

proved that Secret Police organized the whole action in cooperation with traitor Karel Úlehla. Bitterness passed with time, but one can never forget that."

After release, he returned to his old profession as an electrician in the Prague company Tesla Karlín where he worked until his retirement. As a pensioner, he visited an air club in Benešov and flew sports planes until age seventy. He was fully rehabilitated after Velvet Revolution in 1991. As a result, he could meet his former mates again from the 68th Sqn and 310th Sqn during a visit to England. He died at 83 in Prague on May 15, 1992.

(I contacted him as one of the first RAF airmen in 1988, still during communism, and we were in touch until his death. Finally, I managed to organize a meeting between him and my father after 48 years. So much water passed under the bridge. Karel Šeda was a fine, quiet, and friendly man V.F.)

LEST WE FORGET

At the end of this book, we would like to get into a few numbers that are probably unknown to the broader public

In 1940-1945, 2430 Czech airmen were members of the RAF.

During the Battle of France, 19 Czech pilots were killed

After direct contact with the enemy, 236 Czech airmen were killed

Crashes not caused by the enemy killed 213 Czech airmen

In total, 51 Czechs became POWs and 50 survived the war.

During the infamous Battle of Britain, a total of 88 Czech pilots took part

In total, 93 Czechs were drafted in Canada to form a contingent of settled Czech emigrants

In January 1943, a total of 160 soldiers who came from the Czech army fighting in the Middle East joined the RAF

Further, 114 soldiers came from the Government Army, Slovakian Army, and German Army, ran away from forced labor in Germany, escaped from concentration camps in Germany, and all joined the RAF

In 1944 total of 40 airmen voluntarily left the RAF and moved to the Soviet Union to help improve the Russian Air force and contribute to it in fights on the Eastern front, 4 of them got killed

19 Czech women were serving in WAAF

Otto Smik DFC, Jiří Manak DFC, and Frantisek Fajtl DFC were 3 Czechs who were commanders of British squadrons

In total, 765 members of the RAF we sent back to Czechoslovakia after the war.

Altogether 112 Czech airmen received various British military awards such as MBE, DFC, DFM, AFC, DSO

After the war, 168 Czechs were demobbed from RAF in Great Britain

Altogether 40 Czech pilots asked for repatriation

In total, 153 Czechs married in England during the war

The end of war sadly didn't mean the end of victims. 42 Ex-RAF airmen got killed in various accidents in the post-war years.

Many relatives and families of escaped airmen were persecuted by Nazis. In total, 195 families were sent to a detention camp in Svatobořice in Moravia

Most of the former members of the RAF were persecuted by the communist regime after February 1948. They and their families were humbled and kicked out of their jobs, houses, and flats. Their property was often illegally confiscated by the almighty regime and Secret Police without reason or any legal documents. Children were denied further education, and parents could do only hard, inferior jobs.

In total, 117 Czech airmen were sent to jail or killed by the regime in the '50s

After Velvet Revolution in 1989, President Václav Havel declared moral rehabilitation to all members of the RAF who fought during WWII. The massive and lengthy task of finding all living members, relatives, or families of the killed or deceased ones was set. This process had three lines:

1) Rehabilitation of living members in the Czech Republic found 186

2) Rehabilitation of killed found 519

3) Rehabilitation of deceased found 277

4) Rehabilitation of living abroad found 231

5) Rehabilitation of deceased abroad found 199

Total number of rehabilitated 1412

From those 231 found living abroad were in:

Great Britain 153

USA 34

Canada 20

Australia 8

Switzerland 5

New Zealand 3

Austria 3

Spain 2

Peru 1

West Germany 1

Italy 1

In 1999 still, 958 former airmen could not be rehabilitated. It was impossible to find their addresses or even the addresses of their relatives or even the fact that anybody was still alive.

This moral rehabilitation couldn't return lost lives to loved ones or lost and wasted years to those still alive. But it was a human gesture by a human president who was in jail and humbled by the same regime. It couldn't erase the scars in the souls of all concerned, and it certainly couldn't erase the feeling of wrongdoing. It was and still is a black spot in the history of the Czech Republic.

During the writing this book (March 2020), there were four known cases of living Czech ex-RAF airmen. One-Emil Boček is

the only still living Czech pilot on a combat mission during the war. Another - Tomas Lowenstein - from the 311th Squadron was utterly forgotten for 65 years. He didn't participate in any get-togethers, never went public, and didn't ask for rehabilitation. Then, a few years ago, he resurfaced at the age of 90.

It won't take too long before all people actively involved in WWII will die. Then, they disappear and become part of history. But we can't face our future if we don't learn from history.

Let's not forget our history

SOURCES

Formánek Vítek. 2002. Válka v zajetí.

Tonder Ivo and Ladislav Sitenský. 1995. Na nebi i v pekle.

David Zdeněk. 1994. Arnošt Valenta.

Rail Jan and Václav Kolesa. 2004. František Burda-Osudný doprovod nad Brest

Horáková Jana and Jaroslav Popelka. 2001. Zdeněk Škarvada-Keep floating!

Kolesa Václav and Jaroslav Popelka. 2004. Karel Schoř-Od "beranů" Wellingtonu do jáchymovského koncentráku.

Šiška Alois. 2000. KX-B neodpovídá.

Jánský Filip and Jiří Drebota. 1999. Vzpomínky nebeského jezdce.

Jísa Zdeněk 1992. Letci Písecka na bojištích druhé světové války.

Rajlich Jiří 2003. Na nebi hrdého Albionu 5. část.

ABOUT THE AUTHORS

Vítek Formánek (1963) lives in East Bohemia, Czech Republic. He has a deep interest in the Royal Air Force (RAF) since 1988. He befriended dozens and dozens of English ex-Guinea Pigs and POWs and has written 11 books about them. He also has written books about punk rock, speedway, film, autograph collecting, homelessness, and handicapped people, with 23 titles so far. He works as a social worker. Vítek has had over 1500 articles published in the Czech and English media.

Eva Csölleová (1964) lives in Eastern Bohemia, Czech Republic. Along with Vítek, she has co-written 13 books about punk rock, film, the Royal Air Force (RAF), autograph collecting, handicapped people, and homelessness. She is a photographer and many of her photos have been used in her books and publications.

CHAPTER 21: ABOUT THE AUTHORS

Vítek Formánek and Eva Csölleová at the
2022 Zlín Film Festival

Enjoy Additional Sastrugi Press Books

50 Florida Wildlife Hotspots by Moose Henderson Ph.D.

This is a definitive guide to finding where to photograph wildlife in Florida. Follow the guidance of a professional wildlife photographer as he takes you to some of the best places to see wildlife in the Sunshine State.
www.sastrugipress.com/books/50-florida-wildlife-hotspots/

50 Jackson Hole Photography Hotspots by Aaron Linsdau

A guide to the best Jackson Hole photography spots. Learn what locals and insiders know to create the most impressive and iconic photography locations in the United States.
www.sastrugipress.com/books/50-jackson-hole-photography-hotspots/

50 Wildlife Hotspots by Moose Henderson Ph.D.

Find out where to find animals and photograph them in Grand Teton National Park from a professional wildlife photographer. This unique guide shares the secret locations with the best chance at spotting wildlife.
www.sastrugipress.com/books/50-wildlife-hotspots/

 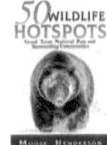

Adventure Expedition One
by Aaron Linsdau M.S. & Terry Williams, M.D.

Create, finance, enjoy, and return safely from your first expedition. Learn the techniques explorers use to achieve their goals and have a good time doing it. Acquire the skills, find the equipment, and learn the planning necessary to pull off an expedition.
www.sastrugipress.com/books/adventure-expedition-one/

Alaska: Illustrated Guide for the Curious
by Nikki Mann and Jeff Wohl

Discover the natural world of Alaska. Find out what the plants and animals are like, how to identify them, and what the environment of Alaska is like.
www.sastrugipress.com/books/alaska-a-guide-for-the-curious/

Along the Sylvan Trail by Julianne Couch

Along the Sylvan Trail dips into the lives of struggling humans as they confront futures that aren't clearly dictated by conventional planning. Share in their quest to find their place in the world.
www.sastrugipress.com/books/along-sylvan-trail/

Antarctic Tears by Aaron Linsdau
Experience the honest story of solo polar exploration. This inspirational true book will make readers both cheer and cry. Coughing up blood and fighting skin-freezing temperatures were only a few of the perils Aaron Linsdau faced.
www.sastrugipress.com/books/antarctic-tears/

Blood Justice by Tim W. James
Two brothers, one a preacher's son, the other an adopted would-be slave, set out in opposite directions to avenge their family's murder only to cross paths in pursuit of the killer.
www.sastrugipress.com/iron-spike-press/blood-justice/

Counterfeit Justice by Tim W. James
Preacher Roger Brinkman takes his crucifix and his Colt to fulfill a promise and help his lawman brother battle thieves, counterfeiters, and murderers in the Old West.
www.sastrugipress.com/iron-spike-press/counterfeit-justice/

 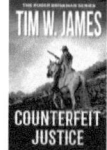

How to Keep Your Feet Warm in the Cold by Aaron Linsdau
Keep your feet warm in cold conditions on chilly adventures with techniques described in this book. Packed with dozens and dozens of ideas, learn how to avoid having cold feet ever again in your outdoor pursuits.
www.sastrugipress.com/books/how-to-keep-your-feet-warm-in-the-cold/

Jackson Hole Hiking Guide by Aaron Linsdau
Find the best hiking trails in Jackson Hole. You'll get maps, GPS coordinates, accurate routes, elevation info, highlights, and dangers. The guide includes easy, challenging, family-friendly, and ADA-accessible trails and hikes.
www.sastrugipress.com/books/jackson-hole-hiking-guide/

Journeys to the Edge by Randall Peeters, Ph.D.
What is it like to climb Mount Everest? Is it possible for you to actually make the ascent? It requires dreaming big and creating a personal vision to climb the mountains in your life. Randall Peeters shares his successes and failures and gives you some directly applicable guidelines on how you can create a vision for your life.
www.sastrugipress.com/books/journeys-to-the-edge/

Lost at Windy Corner by Aaron Linsdau
Windy Corner on Denali has claimed fingers, toes, and even lives. What would make someone brave lethal weather, crevasses, and avalanches to attempt to summit North America's highest mountain? Aaron Linsdau shares the experience of climbing Denali alone and how you can apply the lessons to your life.
www.sastrugipress.com/books/lost-windy-corner/

 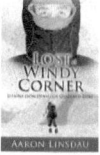

Sleeping Dogs Don't Lie by Michael McCoy
   A young Native American boy is taken from his home after tragedy strikes, grows up in middle America, and through his first real adult summer searches for Wyoming artifacts and attempts single-handedly to solve the mystery behind the murder of his treasured coworker.
   www.sastrugipress.com/books/sleeping-dogs-dont-lie/

So I Said by Gerry Spence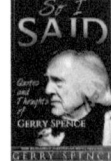
   Venture into the mind of America's most famous lawyer. He shares his thoughts on hope, love, oppression, power, and life. Gain insight from a man who has fought overwhelming power and won from small-town Wyoming.
   www.sastrugipress.com/books/so-i-said/

The Blind Man's Story by J.W. Linsdau
   While on vacation, journalist Beau Larson encounters a blind man high on a forest bluff. This leads him to a brewing war between conservationists and the timber industry, resulting in a mysterious murder.
   www.sastrugipress.com/books/the-blind-mans-story/

The Burqa Cave by Dean Petersen
   Still haunted by Iraq, Tim Ross finds solace teaching high school in Wyoming. That is, until freshman David Jenkins reveals the murder of a lost local girl. Will Tim be able to overcome his demons to stop the murderer?
   www.sastrugipress.com/books/the-burqa-cave/

The Diary of a Dude Wrangler by Struthers Burt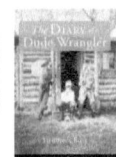
   The dude ranch world of Struthers Burt was a romantic destination in the early twentieth century. He made Jackson Hole a tourist destination. These ranches were and still are popular destinations. Experience the origins of the modern old west.
   www.sastrugipress.com/books/diary-of-a-dude-wrangler/

The Most Crucial Knots to Know by Aaron Linsdau
   Knot tying is a skill everyone can use in daily life. This book shows how to tie over 40 of the most practical knots for virtually any situation. This guide will equip readers with skills that are useful, fun to learn, and will make you look like a confident pro.
   www.sastrugipress.com/books/the-most-crucial-knots-to-know/

The Motivated Amateur's Guide to Winter Camping
by Aaron Linsdau

Winter camping is one of the most satisfying ways to experience the wilderness. It is also the most challenging style of overnighting in the outdoors. Learn 100+ tips from a professional polar explorer on how to winter camp safely and be comfortable in the cold.
www.sastrugipress.com/books/the-motivated-amateurs-guide-to-winter-camping/

Two Friends and a Polar Bear
by Terry Williams, M.D. & Aaron Linsdau

This story of friendship is about two old friends who plan to ski across the Greenland Icecap along the Arctic Circle in hopes of becoming one of the oldest teams to succeed.
www.sastrugipress.com/books/two-friends-and-a-polar-bear/

Shake Yourself Free by Bob Millsap

Learn how to overcome difficult encounters with misfortune, tragedy, and loss. Emotional recovery is a journey requiring a mindset shift. Get this book now and take control of your life.
www.sastrugipress.com/books/shake-yourself-free/

Voices at Twilight by Lori Howe, Ph.D.

Lori Howe invites the reader into the in-between world of past and present in this collection of poems, historical essays, and photographs, all as hauntingly beautiful and austere as the Wyoming landscape they portray.
www.sastrugipress.com/books/voices-at-twilight/

Use your smart device to scan the QR codes to visit website links.

Visit Sastrugi Press on the web at www.sastrugipress.com to purchase the above titles in bulk. They are also available from your local bookstore or online retailers in print, e-book, or audiobook form. Thank you for choosing Sastrugi Press.

www.sastrugipress.com
"Turn the Page Loose"

Author's Note

We hope you enjoyed this book. Please consider giving it a 5-star rating and add a few words about your reading experience at your favorite online retailer.

The link below with a QR code will take you to the book's web page. From there, you can follow links to various online retailers.

Giving our book 5-stars and a short written review about why you enjoyed it will help us immensely.

Thank you!

Vítek Formánek and Eva Csölleová

https://www.sastrugipress.com/books/wasted-lives-of-unsung-heroes/

Use your smart device to scan the QR code for the book's webpage.

www.ingramcontent.com/pod-product-compliance
Lightning Source LLC
Chambersburg PA
CBHW022056160426
43198CB00008B/253